ISAAC ASI[...] foremost wr[...]
An Associate[...]
the Boston Un[...] Medicine,
he has written [...] hundred books,
as well as hund[...]s of articles in publications ranging from *Esquire* to Atomic Energy Commission pamphlets. Famed for his science fiction writing (his three-volume Hugo Award-winning THE FOUNDATION TRILOGY is available in individual Avon editions and as a one-volume Equinox edition), Dr. Asimov is equally acclaimed for such standards of science reportage as THE UNIVERSE, LIFE AND ENERGY, THE SOLAR SYSTEM AND BACK, ASIMOV'S BIOGRAPHICAL ENCYCLOPEDIA OF SCIENCE AND TECHNOLOGY, and ADDING A DIMENSION (all available in Avon editions). His non-science writings include the two-volume ASIMOV'S GUIDE TO SHAKESPEARE, ASIMOV'S ANNOTATED DON JUAN, and ASIMOV'S GUIDE TO THE BIBLE (available in a two-volume Avon edition). Born in Russia, Asimov came to this country with his parents at the age of three, and grew up in Brooklyn. In 1948 he received his Ph.D. in Chemistry at Columbia and then joined the faculty at Boston University, where he works today.

Other Avon books by
Isaac Asimov

THE PLANET
THAT WASN'T

ISAAC ASIMOV

A DISCUS BOOK/PUBLISHED BY AVON BOOKS

The following essays in this volume are reprinted from *The Magazine of Fantasy and Science Fiction*, having appeared in the indicated issues:

Star in the East, December 1974
Thinking About Thinking, January 1975
The Rocketing Dutchmen, February 1975
The Bridge of the Gods, March 1975
The Judo Argument, April 1975
The Planet That Wasn't, May 1975
The Olympian Snows, June 1975
Titanic Surprise, July 1975
The Wicked Witch is Dead, August 1975
The Wrong Turning, September 1975
The Third Liquid, October 1975
Best Foot Backward, November 1975
The Smell of Electricity, December 1975
Silent Victory, January 1976
Change of Air, February 1976
The Nightfall Effect, March 1976
All Gall, April 1976
© 1974, 1975, 1976 by Mercury Press, Inc.

AVON BOOKS
A division of
The Hearst Corporation
959 Eighth Avenue
New York, New York 10019

Copyright © 1976 by Isaac Asimov
Published by arrangement with Doubleday & Company, Inc.
Library of Congress Catalog Card Number: 75-40710
ISBN: 0-380-01813-6

First Discus Printing, December, 1977
Third Printing

DISCUS TRADEMARK REG. U.S. PAT. OFF. AND IN
OTHER COUNTRIES, MARCA REGISTRADA, HECHO EN
U.S.A.

Printed in the U.S.A.

Dedicated to
the memory of James Blish (1921–1975)

Contents

constitution. The liquid hydrogen, with its methane/etc. impurities, rises and falls in a massive circulation that may take a year to rise to the upper zone, and somewhere turn and take another to fall through it again.

If there is fire on Jupiter, it may inhabit those rising

Introduction

The key to my long-continuing series of essays in *The Magazine of Fantasy and Science Fiction* is variety. Partly that is to the credit of my Gentle Editors—Edward L. Ferman at *F & SF* and Cathleen Jordan at Doubleday & Company—who are perfectly content to let me maunder on about any subject that interests me. Partly, also, it is to the credit of my own restless mind.

I am quite likely to talk about the atmosphere one month, space colonics the next, and atherosclerosis the one after. Why not? Not only do I hold my own interest that way, but I pique the readers' curiosity. "What will Isaac be talking about next month?" they wonder.

However, be thou as chaste as ice, as pure as snow, thou shalt not escape calumny.*

Would you believe that last week I received a letter denouncing me for having been on an "astronomy kick" for years and urging me, quite angrily, to get off?

"What is this?" thought I, in confusion. "Have I been writing nothing but astronomy articles for years and haven't noticed?"

I checked and I leave it to you. If you'll consider this book, which has my most recent seventeen essays in it, you will see that there are exactly four articles that might be considered straightforward astronomy—the first four in the book. Some of the rest may have tangential astronomic interest, but nothing more than that. I

---

* That's not one of my own. It was written by an old-time dramatist named William Shakespeare. One of his characters with the improbable name of Hamlet says it to Ophelia—apparently a girlfriend of his.

can only suppose my correspondent had been nipping just a bit too eagerly at the medicinal brandy.

The variety of my essays does present problems, however. For instance, what do I call the collections?

Looking back on it, I can see that I would have saved myself a lot of trouble if I had called them all "Collected Essays" and labeled them Volume 1, Volume 2, Volume 3, and so on.—But how dull!

What I do, instead, is to try to pick some bouncy title that (1) indicates at least vaguely the nature of the contents, and (2) uses a key word that symbolizes science and yet one that I have not used before.

If one of the essays included in the book legitimately has the title I select, so much the better.

I haven't used the word "planet" in any of the titles to my essay collections, and one of the essays in this collection has the title "The Planet That Wasn't." Very good! That's the title of the book.

It sounds like science fiction, I admit, but that's not necessarily a drawback. Any of my books, whatever the title, is likely to be put in the science fiction section, anyway. You can't always count on bookstore employees knowing that I write anything but science fiction.

I once got an agitated phone call from a friend of mine who told me that a book of mine entitled *An Easy Introduction to the Slide Rule*—which was exactly what the title proclaimed it to be—was in with my science fiction at a certain bookstore. I told him to relax, since it is precisely the readers of my science fiction who are most apt to buy and read my nonfiction.

Another problem raised by the variety of essays I write is the matter of arrangement. I have seventeen essays in this book. Which comes first, which second, which third, and why?

I could arrange them in the order they were originally published but that would mean that, by and large, they would be arranged every which way. That sort of charming disorder works fine as long as they appear at monthly intervals in *F & SF*. The reader, with a month between each two successive articles, a month in which he is engaged in all sorts of occupations and preoccupations, has no clear memory of the essay of the month before, unless I remind him. He is therefore ready for anything and follows me anywhere.

The situation is quite different when the essays appear in a book as they do here. Then, you have them all at a whack and the reader is likely to read them all in a fairly short time—weeks, days. There may even be some hardy souls who will read the collection at one prolonged sitting.

Now, you see, the random order is not so good. Having gotten the reader into a certain mood with one essay, I would like to capitalize on that mood for the next, if I can. Therefore, I frequently group my essays by general subject matter.

In this book, however, I have the opportunity to try something else. Let me explain—

I am encouraged by Ed Ferman to write what we call "controversials." Every once in a while he wants me to discuss some issue that, for some reason or another, is a delicate one.

In every case, I uphold the cause of science in as forthright and belligerent a manner as I can possibly manage. Whether I am denouncing flying saucers, or IQ tests, or resistance to space colonization, I do so without thought of compromise or appeasement. Here I stand; I cannot do otherwise.†

This results in a fair amount of mail, which delights good old Ed, so I think I am being lured into doing more and more of these. No, I take back that "lured." The fact is I enjoy the controversials and am delighted at the chance of speaking my mind.

In this book, then, I have decided to arrange my articles in a new way—in the direction of increasing

---

† Again, not one of my own. A preacher named Martin Luther said it—and, for some reason, chose to say it in German, "Hier steh' ich, ich kann nicht anders."

controversiality. We start off with straight science and head toward opinion about science.

This means that if you are in a mood for an argument, you might be tempted to start with the last essay and read forward. Of course, it might then turn out that you agree with me in every particular and that you will have lost your chance at an argument. In that case, I would apologize—

Isaac Asimov
New York City

1

The Planet That Wasn't

I was once asked whether it was at all possible that the ancient Greeks had known about the rings of Saturn. The reason such a question is raised at all comes about as follows—

Saturn is the name of an agricultural deity of the ancient Romans. When the Romans had reached the point where they wanted to match the Greeks in cultural eminence, they decided to equate their own uninteresting deities with the fascinating ones of the imaginative Greeks. They made Saturn correspond with Kronos, the father of Zeus and of the other Olympian gods and goddesses.

The most famous mythical story of Kronos (Saturn) tells of his castration of his father Ouranos (Uranus), whom he then replaced as ruler of the Universe. Very naturally, Kronos feared that his own children might learn by his example and decided to take action to prevent that. Since he was unaware of birth-control methods and was incapable of practicing restraint, he fathered six children (three sons and three daughters) upon his wife, Rhea. Taking action after the fact, he swallowed each child immediately after it was born.

When the sixth, Zeus, was born, Rhea (tired of bearing children for nothing) wrapped a stone in swaddling clothes and let the dim-witted lord of the Universe swallow that. Zeus was raised in secret and when he grew up he managed, by guile, to have Kronos vomit up his swallowed brothers and sisters (still alive!). Zeus and his siblings then went to war against Kronos and *his* siblings (the Titans). After a great ten-year struggle, Zeus defeated Kronos and took over the lordship of the Universe.

Now, then, let's return to the planet which the Greeks had named Kronos, because it moved more slowly against the background of the stars than any other planet and therefore behaved as though it were an older god. Of course, the Romans called it Saturn, and so do we.

Around Saturn are its beautiful rings that we all know about. These rings are in Saturn's equatorial plane, which is tipped to the plane of its orbit by 26.7 degrees. Because of this tipping, we can see the rings at a slant.

The degree of tip is constant with respect to the stars, but not with respect to ourselves. It appears tipped to us in varying amounts depending on where Saturn is in its orbit. At one point in its orbit, Saturn will display its rings tipped downward, so that we see them from above. At the opposite point they are tipped upward, so that we see them from below.

As Saturn revolves in its orbit, the amount of tipping varies smoothly from down to up and back again. Halfway between the down and the up, and then halfway between the up and the down, at two opposite points in Saturn's orbit, the rings are presented to us edge-on. They are so thin that at this time they can't be seen at all, even in a good telescope. Since Saturn revolves about the Sun in just under thirty years, the rings disappear from view every fifteen years.

When Galileo, back in the 1610s, was looking at the sky with his primitive telescope, he turned it on Saturn and found that there was something odd about it. He seemed to see two small bodies, one on either side of Saturn, but couldn't make out what they were. Whenever he returned to Saturn, it was harder to see them until, finally, he saw only the single sphere of Saturn and nothing else.

"What!" growled Galileo, "does Saturn still swallow his children?" and he never looked at the planet again. It was another forty years before the Dutch astronomer Christiaan Huygens, catching the rings as they were tipping further and further (and with a telescope better than Galileo's), worked out what they were.

Could the Greeks, then, in working out their myth of Kronos swallowing his children, have referred to the planet Saturn, its rings, the tilt of its equatorial plane, and its orbital relationship to Earth?

No, I always say to people asking me this question, unless we can't think up some explanation that is simpler and more straightforward. In this case we can—coincidence.

People are entirely too disbelieving of coincidence. They are far too ready to dismiss it and to build arcane structures of extremely rickety substance in order to avoid it. I, on the other hand, see coincidence everywhere as an inevitable consequence of the laws of probability, according to which having no unusual coincidence is far more unusual than any coincidence could possibly be.

And those who see purpose in what is only coincidence don't usually even know the really good coincidences—something I have discussed before.* In this case what about other correspondences between planetary names and Greek mythology? How about the planet that the Greeks named Zeus and the Romans named Jupiter? The planet is named for the chief of the gods and it turns out to be more massive than all the other planets put together. Could it be that the Greeks knew the relative masses of the planets?

The most amazing coincidence of all, however, deals with a planet the Greeks (you would think) had never heard of.

Consider Mercury, the planet closest to the Sun. It has the most eccentric orbit of any known in the nineteenth century. Its orbit is so eccentric that the Sun, at the focus of the orbital ellipse, is markedly off-center.

When Mercury is at that point in its orbit closest to the Sun ("perihelion"), it is only 46 million kilometers

---

* See "Pompey and Circumstance" in *The Left Hand of the Electron* (Doubleday, 1972).

away and is moving in its orbit at a speed of fifty-six kilometers a second. At the opposite point in its orbit, when it is farthest from the Sun ("aphelion"), it is 70 million kilometers away and has, in consequence, slowed down to thirty-seven kilometers a second. The fact that Mercury is sometimes half again as far from the Sun as it is at others, and that it moves half again as quickly at some times than at others, makes it somewhat more difficult to plot its movements accurately than those of the other, more orderly, planets.

This difficulty arises most noticeably in one particular respect—

Since Mercury is closer to the Sun than Earth is, it occasionally gets exactly between Earth and Sun and astronomers can see its dark circle move across the face of the Sun.

Such "transits" of Mercury happen in rather irregular fashion because of the planet's eccentric orbit and because that orbit is tilted by seven degrees to the plane of Earth's orbit. The transits happen only in May or November (with November transits the more common in the ratio of 7 to 3) and at successive intervals of thirteen, seven, ten, and three years.

In the 1700s, transits were watched very eagerly because they were one thing that could not be seen by the unaided eye and yet could be seen very well by the primitive telescopes of the day. Furthermore, the exact times at which the transit started and ended and the exact path it took across the solar disc changed slightly with the place of observation on Earth. From such changes, the distance of Mercury might be calculated and, through that, all the other distances of the Solar System.

It was very astronomically embarrassing, then, that the prediction as to when the transit would take place was sometimes off by as much as an hour. It was a very obvious indication of the limitations of celestial mechanics at the time.

If Mercury and the Sun were all that existed in the Universe, then whatever orbit Mercury followed in circling the Sun, it would follow it exactly in every succeeding revolution. There would be no difficulty in predicting the exact moments of transits.

However, every other body in the Universe also pulls at Mercury, and the pull of the nearby planets—Venus, Earth, Mars, and Jupiter—while very small in comparison to that of the Sun, is large enough to make a difference.

Each separate pull introduces a slight modification in Mercury's orbit (a "perturbation") that must be allowed for by mathematical computations that take into account the exact mass and motion of the object doing the pulling. The resulting set of complications is very simple in theory since it is entirely based on Isaac Newton's law of gravitation, but is very complicated in practice since the computations required are both lengthy and tedious.

Still it had to be done, and more and more careful attempts were made to work out the exact motions of Mercury by taking into account all possible perturbations.

In 1843, a French astronomer, Urbain Jean Joseph Leverrier, published a careful calculation of Mercury's orbit and found that small discrepancies persisted. His calculations, carried out in inordinate detail, showed that after all conceivable perturbations had been taken into account, there remained one small shift that could not be accounted for. The point at which Mercury reached its perihelion moved forward in the direction of its motion just a tiny bit more rapidly than could be accounted for by all the perturbations.

In 1882, the Canadian-American astronomer Simon Newcomb, using better instruments and more observations, corrected Leverrier's figures very slightly. Using this correction, it would seem that each time Mercury circled the Sun, its perihelion was 0.104 seconds of arc farther along than it should be if all perturbations were taken into account.

This isn't much. In one Earth century, the discrepancy would amount to only forty-three seconds of arc. It would take four thousand years for the discrepancy to mount up to the apparent width of our Moon and three million years for it to amount to a complete turn about Mercury's orbit.

But that's enough. If the existence of this forward motion of Mercury's perihelion could not be explained, then there was something wrong with Newton's law of gravitation, and that law had worked out so perfectly in every other way that to have it come a cropper now was not something an astronomer would cheerfully have happen.

In fact, even as Leverrier was working out this discrepancy in Mercury's orbit, the law of gravitation had won its greatest victory ever. And who had been the moving force behind that victory? Why, Leverrier, who else?

The planet Uranus, then the farthest known planet from the Sun, also displayed a small discrepancy in its motions, one that couldn't be accounted for by the gravitational pull of the other planets. There had been suggestions that there might be still another planet, farther from the Sun than Uranus was, and that the gravitational pull of this distant and still unknown planet might account for the otherwise unaccounted-for discrepancy in Uranus's motions.

An English astronomer, John Couch Adams—using the law of gravity as his starting point—had, in 1843, worked out a possible orbit for such a distant planet. The orbit would account for the discrepancy in Uranus's motions and would predict where the unseen planet should be at that time.

Adams's calculations were ignored, but a few months later, Leverrier, working quite independently, came to the same conclusion and was luckier. Leverrier transmitted his calculations to a German astronomer, Johann Gottfried Galle, who happened to have a new star map of the region of the heavens in which Leverrier said there was an unknown planet. On September 23, 1846, Galle began his search and, in a matter of hours, located the planet, which we now call Neptune.

After a victory like that, no one (and Leverrier least of all) wanted to question the law of gravity. The discrepancy in Mercury's orbital motions had to be the result of some gravitational pull that wasn't being taken into account.

For instance, a planet's mass is most easily calculated if it has satellites moving around it at a certain distance and with a certain period. The distance-period combination depends upon the planetary mass, which can thus be calculated quite precisely. Venus, however, has no satellites. Its mass could only be determined fuzzily, therefore, and it might be that it was actually ten per cent more massive than the astronomers of the mid-nineteenth century had thought. If it were, that additional mass, and the additional gravitational pull originating from it, would just account for Mercury's motion.

The trouble is that if Venus were that much more massive than was supposed, that extra mass would also effect the orbit of its other neighbor, Earth—and disturb it in a way that is not actually observed. Setting Mercury to rights at the cost of upsetting Earth is no bargain, and Leverrier eliminated the Venus solution.

Leverrier needed some massive body that was near Mercury but not too disturbingly near any other planet, and by 1859 he suggested that the gravitational source had to come from the far side of Mercury. There had to be a planet inside Mercury's orbit, close enough to Mercury to account for the extra motion of its perihelion, but far enough from the planets farther out from the Sun to leave them substantially alone.

Leverrier gave to the suggested intra-Mercurial planet the name Vulcan. This was the Roman equivalent of the Greek god Hephaistos, who presided over the forge as the divine smith. A planet that was forever hovering near the celestial fire of the Sun would be more appropriately named in this fashion.

If an intra-Mercurial planet existed, however, why was it that it had never been seen? This isn't a hard question to answer, actually. As seen from Earth, any

19

body that was closer to the Sun than Mercury is would always be in the neighborhood of the Sun, and seeing it would be very difficult indeed.

In fact, there would only be two times when it would be easy to see Vulcan. The first would be on the occasion of a total solar eclipse, when the sky in the immediate neighborhood of the Sun is darkened and when any object that is always in the immediate neighborhood of the Sun could be seen with an ease that would, at other times, be impossible.

In one way, this offers an easy out, since astronomers can pinpoint the times and places at which total solar eclipses would take place and be ready for observations then. On the other hand, eclipses do not occur frequently, usually involve a large amount of traveling, and last only a few minutes.

What about the second occasion for easy viewing of Vulcan? That would be whenever Vulcan passes directly between Earth and Sun in a transit. Its body would then appear like a dark circle on the Sun's orb, moving rapidly from west to east in a straight line.

Transits should be more common than eclipses, be visible over larger areas for longer times, and give a far better indication of the exact orbit of Vulcan— which could then be used to predict future transits, during which further investigations could be made and the properties of the planet worked out.

On the other hand, the time of transit can't be predicted surely until the orbit of Vulcan is accurately known, and that can't be accurately known until the planet is sighted and followed for a while. Therefore, the first sighting would have to be made by accident.

Or had that first sighting already been made? Such a thing was possible, and even likely. The planet Uranus had been seen on a score of occasions prior to its discovery by William Herschel. The first astronomer royal of Great Britain, John Flamsteed, had seen it a century before its discovery, had considered it an ordinary star, and had listed it as "34 Tauri." Herschel's discovery did not consist in seeing Uranus for the first time, but in recognizing it as a planet for the first time.

Once Leverrier made his suggestion (and the discoverer of Neptune carried prestige at the time), astronomers began searching for possible previous sightings of strange objects that would now be recognized as Vulcan.

Something showed up at once. A French amateur astronomer, Dr. Lescarbault, announced to Leverrier that in 1845 he had observed a dark object against the Sun which he had paid little attention to at the time, but which now he felt must have been Vulcan.

Leverrier studied this report in great excitement, and from it he estimated that Vulcan was a body circling the Sun at an average distance of 21 million kilometers, a little over a third of Mercury's distance. This meant its period of revolution would be about 19.7 days.

At that distance, it would never be more than eight degrees from the Sun. This meant that the only time Vulcan would be seen in the sky in the absence of the Sun would be during, at most, the half-hour period before sunrise or the half-hour period after sunset (alternately, and at ten-day intervals). This period is one of bright twilight, and viewing would be difficult, so that it was not surprising that Vulcan had avoided detection so long.

From Lescarbault's description, Leverrier also estimated the diameter of Vulcan to be about two thousand kilometers, or only a little over half the diameter of our Moon. Assuming the composition of Vulcan to be about that of Mercury, it would have a mass about one-seventeenth that of Mercury or one-fourth that of the Moon. This is not a large enough mass to account for all of the advance of Mercury's perihelion, but perhaps Vulcan might be only the largest of a kind of asteroidal grouping within Mercury's orbit.

On the basis of Lescarbault's data, Leverrier calculated the times at which future transits ought to take place, and astronomers began watching the Sun on those occasions, as well as the neighborhood of the Sun whenever there were eclipses.

Unfortunately, there were no clear-cut evidences of Vulcan being where it was supposed to be on predicted occasions. There continued to be additional reports as someone claimed to have seen Vulcan from time to time. In each case, though, it meant a new orbit had to be calculated, and new transits had to be predicted—and then these, too, led to nothing clear-cut. It became more and more difficult to calculate orbits that included all the sightings, and none of them successfully predicted future transits.

The whole thing became a controversy, with some astronomers insisting that Vulcan existed and others denying it.

Leverrier died in 1877. He was a firm believer in the existence of Vulcan to the end, and he missed by one year the biggest Vulcan flurry. In 1878, the path of a solar eclipse was to pass over the western United States and American astronomers girded themselves for a mass search for Vulcan.

Most of the observers saw nothing, but two astronomers of impressive credentials, James Craig Watson and Lewis Swift, reported sightings that seemed to be Vulcan. From the reports, it seemed that Vulcan was about 650 kilometers in diameter and only one fortieth as bright as Mercury. This was scarcely satisfactory, since it was only the size of a large asteroid and could not account for much of the motion of Mercury's perihelion, but it was something.

And yet even that something came under attack. The accuracy of the figures reported for the location of the object was disputed and no orbit could be calculated from which further sightings could be made.

As the nineteenth century closed, photography was coming into its own. There was no more necessity to make feverish measurements before the eclipse was over, or to try to make out clearly what was going on across the face of the Sun before it was all done with. You took photographs and studied them at leisure.

In 1900, after ten years of photography, the American astronomer Edward Charles Pickering announced there could be on intra-Mercurial body that was brighter than the fourth magnitude.

In 1909, the American astronomer William Wallace Campbell went further, and stated categorically that there was nothing inside Mercury's orbit that was brighter than the eighth magnitude. That meant that nothing was there that was larger than forty-eight kilometers in diameter. It would take a million bodies of that size to account for the movement of Mercury's perihelion.†

With that, hope for the existence of Vulcan flickered nearly to extinction. Yet Mercury's perihelion *did* move. If Newton's law of gravitation was correct (and no other reason for supposing its incorrectness had arisen in all the time since Newton) there had to be some sort of gravitational pull from inside Mercury's orbit.

And, of course, there was, but it originated in a totally different way from that which anyone had imagined. In 1915, Albert Einstein explained the matter in his General Theory of Relativity.

Einstein's view of gravitation was an extension of Newton's—one that simplified itself to the Newtonian version under most conditions, but remained different, and better, under extreme conditions. Mercury's presence so close to the Sun's overwhelming presence was an example of the extreme condition that Einstein could account for and Newton not.

Here's one way of doing it. By Einstein's relativistic view of the Universe, mass and energy are equivalent, with a small quantity of mass equal to a large quantity of energy in accordance with the equation $e = mc^2$.

The Sun's enormous gravitational field represents a large quantity of energy and this is equivalent to a certain, much smaller, quantity of mass. Since all mass gives rise to a gravitational field, the Sun's gravitational

---

† This is correct as far as we know. To this day, the only objects known to have approached the Sun more closely than Mercury does have been an occasional comet of negligible mass and the asteroid Icarus, which is only a kilometer or two across.

field, when viewed as mass, must give rise to a much smaller gravitational field of its own.

It is this second-order pull, the small gravitational pull of the mass-equivalent of the large gravitational pull of the Sun, that represents the additional mass and the additional pull from within Mercury's orbit. Einstein's calculations showed that this effect just accounts for the motion of Mercury's perihelion, and accounted further for much smaller motions of the perihelia of planets farther out.

After this, neither Vulcan nor any other Newtonian mass was needed. Vulcan was hurled from the astronomical sky forever.

Now to get back to coincidences—and a much more astonishing one than that which connects Kronos's swallowing of his children with the rings of Saturn.

Vulcan, you will remember, is the equivalent of the Greek Hephaistos, and the most famous myth involving Hephaistos goes as follows—

Hephaistos, the son of Zeus and Hera, at one time took Hera's side when Zeus was punishing her for rebellion. Zeus, furious at Hephaistos's interference heaved him out of heaven. Hephaistos fell to Earth and broke both his legs. Though he was immortal and could not die, the laming was permanent.

Isn't it strange, then, that the planet Vulcan (Hephaistos) was also hurled from the sky. It couldn't die, in the sense that the mass which supplied the additional gravitational pull had to be there, come what may. It was lamed, however, in the sense that it was not the kind of mass that we are used to, not mass in the form of planetary accumulations of matter. It was the mass-equivalent, instead, of a large energy field.

You are not impressed by the coincidence? Well, let's carry it further.

You remember that in the myth about Kronos swallowing his children, Zeus was saved when his mother substituted a stone in the swaddling clothes. With a stone serving as a substitute for Zeus, you would surely be willing to allow the phrase "a stone" to be considered the equivalent of "Zeus."

Very well, then, who flung Hephaistos (the mythical Vulcan) from the heavens? Zeus!

And who flung the planetary Vulcan from the heavens? Einstein!

And what does *ein stein* mean in Einstein's native German? "A stone"!

I rest my case.

We can say that the Greeks must have foreseen the whole Vulcanian imbroglio right down to the name of the man who solved it.—Or we can say that coincidences can be enormously amazing—and enormously meaningless.

2

The Olympian Snows

I am very concerned about the titles of these essays. When I don't have a good title, I have trouble beginning. Sometimes when I think of a very good title, I deliberately invent an essay to fit around it. Since this essay is the two-hundredth in the *F & SF* series, I felt it necessary to pick a significant subject and build it around a particularly good title—poetic, witty, surprising, *something*.

As for a topic, it occurred to me that there is nothing so dramatic for a science-fictionish person such as myself as the Martian canals. Virtually no writer of twentieth-century science fiction has failed to mention them.

At once it occurred to me, for reasons that will become apparent as I go along, that "The Snows of Olympus" would be a perfect title. I was very pleased with myself and made up my mind that as soon as the appropriate time came I would prepare an essay with that title on that subject.

Then, a few days later, when I was whiling away some moments at a newsstand, I suddenly became aware of the name of my good friend Arthur C. Clarke on the cover of the current *Playboy,* though I don't remember how I came to be looking in that direction. Interested to see what my dear friend Arthur might have to say, I steered austerely past acres of female skin and reached the indicated page.

—And do you know what Arthur had there? A very brief discussion of Mars, and the title he had tacked on was "The Snows of Olympus." I'm probably the only person in history who gasped and choked and jumped ıp and down while staring at a page in *Playboy* that had no trace of womankind upon it.

I had to think quickly and I did. The next time I meet my rotten friend Arthur, I intend to choke him and beat his head against the wall since it's clear to me he did it on purpose. And meanwhile, I quickly changed the title of my article into something completely different, as you will already have noticed.

And now, to work—

The first telescopic discovery made when Galileo turned his initial spyglass on the sky in 1609 was that of the mountains and craters on the Moon. Galileo himself was able to make the first crude drawing of the Moon's surface, and with better and better telescopes built, other astronomers drew better, more detailed, and more accurate maps of the Moon's surface.

It might have seemed at that time that if only telescopes could be made steadily bigger and better, astronomers could, in similar fashion, map all the other worlds of the Solar System.

Alas, it proved not to be so. The great outer planets —Jupiter, Saturn, Uranus, and Neptune—are perpetually cloud-covered, and all we can map are the cloud bands of Jupiter and Saturn. As for the smaller bodies of the outer Solar System—asteroids, satellites, and the like—no telescope was ever built (or is ever likely to be built on Earth's surface) that could make them out well enough to give any surface detail, even where no obscuring atmosphere exists.

That leaves us with the objects of the inner Solar System, other than the Earth and the Moon, as possible map-targets. There are only five of these. Working outward from the Sun, there are Mercury, Venus, Mars, and the two Martian moons, Phobos and Deimos.

Of these five, Phobos and Deimos are too small to show up as anything but dots of light in even the best telescopes, and Venus is perpetually cloud-covered, and with featureless cloud at that. Mercury lacks an atmosphere and exposes its bare surface, but when it is most easily studied, it is 110 million kilometers away, is visible as a fat crescent with most of its surface dark, and is too close to the Sun to make for easy viewing. All that can be seen of the Mercurian surface from Earth-based telescopes is just vague splotches that have never amounted to anything much.

That leaves Mars as the *only* object, other than Earth and the Moon, that mankind could possibly have mapped prior to the Space Age.

Earth's average distance from the Sun is 150 million kilometers, while Mars's average distance is 228 million kilometers. If both moved around the Sun in perfectly circular orbits, then every time Earth passed Mars ("opposition") the two planets would be 78 million kilometers apart.

The orbits are not circular, however, but somewhat elliptical so that they are closer together in some places than in others. The two orbits may be as far apart as 99 million kilometers, or as close together as 56 million kilometers.

It is always best to observe Mars at opposition, when it is closer to us than it will be for months before and months afterward, and when it shines high in the midnight sky, with its entire surface facing us brightly sunlit. When the opposition takes place at times when the two planets are moving through those portions of their orbits that are relatively close together, so much the better. At the closest opposition, Mars is only about 150 times as far away as the Moon and no other sizable body but the cloud-covered Venus ever gets that close to Earth.

The first close opposition after telescopes had become a common adjunct of astronomy was in 1638, and in that year the Italian astronomer Francesco Fontana made the first attempt to draw what he saw when he looked at Mars. Since he didn't see much, we can only record it as a first attempt and pass on.

The first astronomer to see something on Mars that eventually came to be accepted as a real feature of the surface was the Dutchman Christiaan Huygens. On November 28, 1659, he drew a picture of Mars that included a V-shaped dark spot in the equatorial region. It continued to appear in every pictorialization of the Martian surface thereafter.

On August 13, 1672, Huygens went on to draw another map, on which he indicated an icecap for the first time.

Both Huygens and the Italian-French astronomer Giovanni Domenico Cassini tried to note the changes in position of the various vague spots they saw on the Martian surface from night to night, and to make use of such changes to determine the planetary rotation period.

In 1664, Cassini found the Martian rotation to have a period of 24 hours and 40 minutes. This is only 2.6 minutes faster than the figure now accepted and is certainly not bad for a first try.

As observation of Mars continued, its similarities to Earth strengthened. Not only was the Martian day very similar in length to the Earth day, but the inclination of the Martian axis to its plane of revolution about the Sun (25.2 degrees) was very similar to Earth's 23.5 degrees. That meant that Mars had seasons very like those of Earth, except that each was almost twice as long as Earth's and, on the whole, considerably colder.

The German-English astronomer William Herschel, studying Mars in the 1770s and 1780s, noted the presence of an atmosphere on Mars and detected color changes with the seasons.

All of this was important in connection with the problem of life on other worlds.

In early modern times, astronomers had a tendency to assume that all worlds were inhabited, if only because it seemed sacrilegious to suppose that God would create a world and let it be wasted. Yet everything astronomers learned about the worlds of the Solar System went against this supposition. The world that was closest and best-known, the Moon, plainly had neither air nor water and couldn't possibly support life after Earth's fashion. And if it was a dead world, surely others might be, too.

Naturally, this was disappointing—and by the world outside it was ignored. The average layman continued to assume life on all planets, and so did the science fiction writer. (In one of my first published stories, "The Callistan Menace," I calmly gave the Jovian satellite Callisto a native life of its own.)

Astronomers, however, could not console themselves with any such romantic escapes. The Solar System increasingly seemed a collection of worlds that were, for the most part, dead—and the more that seemed to be so, the more astronomers found themselves attracted to Mars, which, with its axial tilt, its icecaps, its color changes, seemed so Earthlike, and therefore so alive.

In 1830, two German astronomers, Wilhelm Beer (a brother of the composer Giacomo Meyerbeer) and Johann Heinrich von Mädler, studied the surface of Mars during a close opposition and produced the first drawings that were recognizable maps of the planet.

Until then, the vague dark and light markings had, for the most part, seemed so vague that observers thought they were cloud formations or patches of mist. Beer and Mädler were the first, however, to determine that some dark and light markings were reasonably permanent, and it was these that they tried to draw.

The map wasn't a very good one by later standards, but they were the first to set up a system of latitude and longitude similar to that on Earth. The lines of latitude, based on the equator and the poles, were simple to define, but the lines of longitude had to be marked off from some arbitrary zero mark. This, Beer and Mädler placed at a small, round, dark marking they saw particularly clearly, and that standard has been modified only slightly since.

Other astronomers in the decades that followed also tried to draw maps. One of them was an English astronomer, Richard Anthony Proctor, who drew a map of Mars in 1867 and was so confident of his results that he decided to name the various features. He called the dark areas oceans, seas, and straits, while the light areas he called continents and lands. He named all the markings after astronomers, living and dead.

The system had worked for the Moon, but Proctor favored English astronomers so heavily that French and German astronomers were bitterly offended and the system was *not* accepted.

Then came 1877, when Mars was to move into opposition at virtually the minimum possible distance. Astronomers, using the best instruments they could, were at hand. One of them was the American astronomer Asaph Hall, who discovered Mars's two tiny satellites during this opposition—but that is another story.

Another was the Italian astronomer Giovanni Virginio Schiaparelli, who, as a result of his observations, was able to draw the first modern map of Mars, one that lasted, with minor modifications, for nearly a century.

More important still, Schiaparelli worked out a new system for naming the Martian features, one that was far more successful than Proctor's and which, in fact, is still used today.

For one thing, Schiaparelli avoided national rivalries by using Latin exclusively, and for another he made use of Mediterranean place names taken from ancient history, mythology, and the Bible. Thus, the dark feature that had first been spotted by Huygens, Schiaparelli named Syrtis Major ("great bog"), for he still assumed, as did everyone else, that the dark markings were water and the light markings land.

Ever since, various Martian markings have received romantic and sonorous Latin names. A light spot located at a Martian longitude of 135°, some twenty degrees north of the Martian equator, was named Nix Olympica, which I choose to translate as "the Olympian Snows."

Like Proctor, Schiaparelli noted narrow, dark markings which crossed the light regions and connected larger dark markings at either end. Proctor had called them "straits" and Schiaparelli called them "channels." Schiaparelli gave the various channels the names of rivers.

Four of them, for instance, were Gehon, Hiddekel, Euphrates, and Phison, from the four rivers in the Garden of Eden. There were Lethes and Nepenthes from the rivers of Hades, and there were Orontes and Nilus from real geography. There was, in all this, no suggestion of anything but natural waterways.

In calling them "channels," however, Schiaparelli used the Italian word *canali,* which (rather naturally) was translated into the English word "canals." Whereas, in English, a channel is a natural waterway, a canal is a man-made one, and that made an enormous difference.

As soon as men began to talk about the "canals of Mars," all the longing to have the worlds of the Universe inhabited and all the particular feelings that Mars, at least, was an Earthlike world came to a head. Mars seemed not only to be inhabited, but it had to have a high civilization capable of irrigating the entire planet with gigantic engineering works.

It became easy, in fact, to work up a very romantic story about Mars. It was a small world, with only one-tenth the mass of Earth and only two-fifths Earth's surface gravity. Mars might hold on to its water so feebly that, little by little, water escaped into space so that Mars grew drier and drier by very slow stages.

Fighting against this gradual desiccation was a gallant, if aging, civilization, making what use it could of the icecaps, the planet's last reservoir of water.

With more and more astronomers eagerly looking at the canals, more and more dramatic phenomena were reported. Individual canals were found to double on occasion. Where canals crossed there were little round areas of darkness, which, in 1892, the American astronomer William Henry Pickering suggested be called "oases."

It was in 1893, however, that the matter of the Martian canals began to come to full flower, for in that year the American astronomer Percival Lowell grew interested.

Lowell, scion of an aristocratic Boston family, had wealth enough to humor his whims, and he built an excellent observatory in the mile-high dry desert air of Flagstaff, Arizona. There he devoted himself to a fifteen-year study of the surface of Mars.

He drew more and more elaborate maps, showing more and more canals, until finally he was able to draw five hundred of them. No one else could see anything like the detail Lowell could, but that did not disturb Lowell. He maintained that other astronomers had poorer eyes and poorer instruments and were watching Mars in poorer climates.

What's more, Lowell insisted that the canals *were* artificial and that Mars *was* the home of an advanced civilization. He presented this view to the general public for the first time in a book named *Mars,* published in 1895.

The general public is, of course, always ready to accept the dramatic, and the Lowellian view was widely hailed by many. Among the enthusiasts was the English writer Herbert George Wells.

In 1898, Wells published *The War of the Worlds.* Following the Lowellian view, Wells pictured Mars as a dying world. Its leaders decided that to remain on Mars was slow suicide and that they must therefore migrate to flourishing, watery Earth. The Martian ships landed on Earth (all of them in England, for some reason, though Wells nowhere indicates he finds this strange) and proceeded to take over the island as brutally and indifferently as we would take one over that was inhabited only by rabbits. The Martians were defeated only when they fell prey to Earth's decay germs, against which they had no defense.

The book, as far as I know, was the first tale of interplanetary warfare ever written and was even more influential than Lowell's book in convincing nonastronomers that there was intelligent life on Mars.

Among astronomers, Lowell's views were not generally accepted. Many of them, including some of the best observers, simply did not see the canals. An Italian astronomer, Vincenzo Cerulli, held that the canals were an optical illusion. There were, he claimed, irregular

33

patches on the Martian surface, patches that were just at the limit of vision. The eye, straining to see them, ran them together into straight lines.

The French astronomer Eugenios Marie Antoniadi made maps of Mars, beginning in 1909, that were superior to Schiaparelli's and he saw no canals; he saw irregular spots as Cerulli had suggested.

And yet many astronomers *did* see the canals, and there seemed no way of settling the matter definitely. None of the advances after Schiaparelli's time seemed to help. New large telescopes were built, and each in turn was eagerly turned on Mars, and each had to fall back defeated. The very large telescopes greatly magnified the image of Mars, but they also magnified the distorting effect of temperature changes in the atmosphere. Although the large telescopes were excellent for studying deep space, they were not so good for studying the nearby planets, at least as long as they had to work from the bottom of an ocean of air.

Nor could the new technique of photography do any good. Photographs of the planets were never as clear as the seeing one could get by way of the eye at the telescope. For one thing, the photographic plates were themselves grainy and that introduced an unavoidable fuzz. For another, the plates required a time exposure and that gave an opportunity for atmospheric imperfections to obscure detail. With the eye, you could get brief flashes of intimate detail at moments when the air was absolutely clear, detail that you could never get in photographs.

Right down to 1965, then, one could still argue over whether there were canals on Mars or not.

As the twentieth century progressed, however, it began to seem less and less likely that the canals, even if they existed, could be the product of an advanced race of intelligent beings that were *now* living on Mars, for as studies continued, the Martian environment began to seem less and less hospitable.

The Martian atmosphere, it turned out, was thinner

than expected, and, what's more, it contained no oxygen at all; it was nothing more than a whiff of carbon dioxide and, possibly, nitrogen.

Again, Mars was even drier than had been expected. The planet had no lakes or seas or bogs, despite the use of Latin words signifying such features. There was no snow, it seemed quite certain, in the area of "the Olympian Snows." The polar icecaps seemed to be the only water of account on the planet and they might be only a few inches thick. For that matter, they might not even be water. There seemed increasing reason to think they were made up of frozen carbon dioxide.

Under these conditions, the canals, if they existed, were useless. Perhaps they might have been useful once, when Mars was milder and had more water and air, but when might *that* have been—if ever?

Yet, despite everything, some astronomers *did* see canals, and most people *did* believe in them.

Nothing more could be done until some view of Mars could be taken under conditions better than those possible on the Earth's surface. Instruments simply had to be sent to the neighborhood of Mars.

On November 28, 1964, a mighty step was taken in this direction when a Mars probe, Mariner 4, was launched. In 1965, it sent back some twenty photographs taken from a distance of 9,500 kilometers above the Martian surface. The photographs showed no signs of canals, no signs of great engineering achievements, no signs of intelligent life. What the photographs *did* show was a Martian surface littered with craters, very much like that of the Moon.

Other data sent back by Mariner 4 seemed to show that the Martian atmosphere was even thinner than the most pessimistic estimate, and the Martian environment more hostile.

On May 30, 1971, another Mars probe, Mariner 9, was launched and sent out toward the planet. On November 14, 1971, it was placed in orbit about 1,600 kilometers above the surface of Mars. This was not a matter of just flying by and catching what photographs it could; Mariner 9 was intended to circle Mars indefinitely and to take photographs for an extended period, eventually (if all went well) mapping the entire surface.

While Mariner 9 was on its way to Mars, a dust storm broke out on the planet and continued for months, obscuring the surface of the world completely. Mariner 9 had to wait. By the end of December 1971 the dust storm subsided, and on January 2, 1972, Mariner 9 began to take its photographs. Eventually, the entire planet was indeed mapped, and it was quickly apparent that the limited sections photographed on earlier missions had not, after all, been representative of the planet as a whole. There were, it is true, large areas that were heavily cratered and seemed Moonlike in nature, but these were largely confined to one hemisphere of the planet. The other hemisphere was like nothing on the Moon, or on the Earth, either.

The most startling feature turned out to be Nix Olympica. There were no snows, of course, but it was much more than the rather unimpressive Mount Olympus of the Greeks. Nix Olympica, the Olympian Snows, was a volcano, a giant volcano, five hundred kilometers wide at the base and therefore twice as wide as the largest volcano on Earth—the one that makes up the island of Hawaii. The crater at the top is sixty-five kilometers wide. Smaller Martian volcanoes were noted in the vicinity of Nix Olympica. Mars was alive—but not in the Lowellian sense.

Southeast of the volcanoes is a system of Martian canyons which also dwarfs anything on Earth. It stretches across a distance equal to the full breadth of the United States; the canyons are up to four times as deep as the Grand Canyon and up to six times as wide —but they, by themselves, could not account for the canals of Lowell.

In fact, there were no canals. With the whole surface of Mars photographed in meticulous detail, there is nothing that can represent what Schiaparelli, Lowell, and a number of others thought they had seen. It *was* optical illusion after all, and the Martian canals which had existed in the minds of men, and in uncounted numbers of science fiction stories (including mine) came to the end of their century-long life.

And Mars became the third world to be mapped in detail.

In addition, the Mars probes caught sight of the little satellites of Mars, irregular potato-shaped objects with craters on them almost as large as they were themselves.

In 1974, a Mercury probe mapped virtually the entire surface of that smallest and Sun-nearest planet, and it turned out to be another heavily cratered world. It looked like a rather more finely stippled Moon, since the individual craters are smaller in comparison with Mercury's size, which is distinctly larger than that of the Moon.

Of all the permanent* worlds of the inner Solar System then, only Venus remains unmapped in detail, since it alone has a surface obscured by clouds. (Earth's is obscured by clouds, too, but we are *under* Earth's clouds.)

All is not lost, though. Venus has been observed by means of radar, which can penetrate the cloud layer, strike the ground, and be reflected from it. From changes in the nature of the radar beam after reflection, some conclusions can be made as to the nature of the ground, and mountain ranges have been roughly plotted on the Venerean surface.

And what about the vast reaches beyond the orbit of Mars?

Pioneer 10, a Jupiter probe that passed that planet in December 1973, sent back among its data a picture of Ganymede, Jupiter's largest satellite. With a mass twice that of our Moon, Ganymede is the most massive satellite in the Solar System.

---

* By using this word, I mean to exclude asteroids, meteoroids, and comets that wander into the inner Solar System at one end of elongated orbits.

The picture is very fuzzy indeed, but until now all that has ever been seen of Ganymede by any instrument has been a point of light or, at best, a small featureless disc. This new picture, then, represents an enormous advance.

The picture seems to show something equivalent to a large lunar sea ("mare") about Ganymere's north pole, and a smaller one near its equator. There are also signs of large craters.

Pioneer 11, which is on its way to Jupiter as I write, may tell us more, and I suspect we are in for more surprises.

Ganymede and Callisto, the two outer ones of Jupiter's four large satellites, have such low densities that it is expected they are made up largely of water ice and ammonia ice. The Olympian Snows that we lost on Mars may reappear here, then, in enormous quantity.†

---

† Since this article first appeared, Nix Olympica has been renamed "Olympus Mons" (Mount Olympus), and Pioneer 11 has taken photographs which show icecaps on Callisto.

3

Titanic Surprise

I keep watching, with more or less apprehension, for scientific discoveries that may completely knock out some article I have previously written. It happens every once in a while and though I should be, and *am,* delighted to see scientific advance convert the more speculatively wrong into the less speculatively more likely right, I am also human enough to mourn the dead article.

Well, mourn with me! Back in the May 1962 *F & SF,* I wrote an article entitled "By Jove," which eventually appeared in my essay collection *View from a Height* (Doubleday, 1963). In it I followed the speculations of Carl Sagan to the effect that the greenhouse effect might give Jupiter a comfortable temperature with a dense atmosphere and a vast mild ocean, both containing just the kind of compounds that would easily develop into life structures. I even calculated that the mass of living matter in Jupiter's oceans might be so great as to be equal to one-eighth the total mass of our Moon.

Alas! From the data sent back by the Jupiter probe Pioneer 10, it looks as though the vision of a comfortable Jupiter is wrong. The planet is, essentially, an enormous drop of liquid hydrogen at more than white-hot temperature. Merely one thousand kilometers below the frigid far-below-zero of the cloud layers, the temperature is already 3,600° C, and that temperature continues to rise to some 54,000° C at the center.

Liquid hydrogen boils at —253° C under Earthly conditions, just twenty degrees above absolute zero;

but Jovian pressures keep it liquid at temperatures well above the surface temperature of the Sun.

We can still imagine Jovian life, of course. As the temperature rises from the frigid clouds downward, it must pass through a level where the temperature is Earth-comfortable. The liquid hydrogen, with its ammonia/methane/etc. impurities, rises and falls in a slow, majestic circulation and may take a year to rise through the tepid zone and somewhere turn and take another year to fall through it again.

If there is life on Jupiter, it may inhabit those rising and falling columns, switching from rise to fall when the temperature gets too low and from fall to rise when it gets too high.

Still, the mood has come upon me to seek another world for my ebullient interest in unlikely life-homes, so I think I'll go over the worlds of the Solar System with some care, dividing them by mass range. For the sake of neatness, I will use a rising scale of ten—1, 10, 100, 1,000, 10,000, and 100,000—for dividing lines. To achieve interesting results I will set the mass of our dear old Moon (73,500,000,000,000,000,000,-000 kilograms) equal to 1.

Let us start at the upper end of the scale and consider objects with masses of more than 100,000 M (that is, more than 100,000 times the mass of the Moon).

The only object of this sort in the Solar System is the Sun, which has a mass of 27,000,000 M. It is, of course, a star and is gaseous throughout. For the most part the gas that composes it is of the type we know on Earth except for the fact of its enormously high temperature. Toward the center of the Sun, the temperature is high enough to break down the atoms and produce a nuclear gas.

White dwarf stars are mostly nuclear gas, neutron stars are a sort of nuclear solid, and black holes are who-knows-what; but for the most part, the stars are gas of the variety we now know as "plasma," because the chipping effects of high temperature produce electrically charged atomic fragments.

Astronomers are agreed that any mass of matter higher than a certain critical amount ends as a star, once it compresses sufficiently under the pull of its own gravitational field. If the mass is high enough, the pressures and temperatures at the center will reach the ignition point of nuclear fusion and that will turn the object into a hot gas.

Exactly what that star-making critical amount of mass may be cannot be said very precisely because, for one thing, it varies somewhat with the properties of the mass. Still, an object with only a tenth the mass of the Sun would still be a star, albeit a "red dwarf," producing only enough heat to make the surface red hot.

Objects smaller still might be gaseous "infrared dwarfs," not quite hot enough to shine visibly. None of these have been unequivocally observed as far as I know, but that is not surprising. They must be so small and deliver so little energy that detection would be difficult indeed.

Yet perhaps we have done so without quite realizing it. Consider that Jupiter is apparently almost massive enough to fulfill the requirement. It radiates three times as much energy into space as it receives from the Sun and this may just possibly be due to very small quantities of fusion at its center—fusion that may help keep the ball of liquid as hot as it is.

If Jupiter, then, were a little larger, more fusion might take place—enough to make the mass a dense gas that would be markedly warm if not quite red hot at the surface. But do we know any object larger (but not very much larger) than Jupiter?

Yes, we do. The faint, rather nearby star 61 Cygni is actually a binary star, with the individuals termed 61 Cygni A and 61 Cygni B. In 1943, the Dutch-American astronomer Peter van de Kamp reported one of those stars to be wobbling slightly and deduced the gravitational effect of a dark companion, 61 Cygni C, a planet about eight times as massive as Jupiter. If so,

41

its mass is about 200,000 M, and it is my guess that, if this is correct, it is an infrared dwarf star.

Let's pass on, though, to the next stage, that between 100,000 M and 10,000 M.

In this range there falls only one known object—the planet Jupiter, which has a mass of 26,000 M. Even if it is large enough to start a few fugitive fusion reactions at its center, the energy generated in this fusion is not enough to render it gaseous—so it is a liquid body. We might call it a substar, perhaps, rather than a giant planet.

In the range between 10,000 M and 1,000 M, there fall three known bodies:

Saturn	7,750 M
Neptune	1,400 M
Uranus	1,200 M

Saturn's density is known to be only about half that of Jupiter. The easiest way of explaining that is to suppose it to be partly gas. Its lesser mass and, therefore, its less intense gravitational field can, perhaps, not compress its hydrogen as tightly and allows much of it to evaporate as a gas.

Uranus and Neptune are roughly as dense as Jupiter. Their lower temperatures may allow more of their structure to be liquid even though their gravitation fields are considerably less intense than those of the two larger planets. Yet there may be substantial quantities of gas there. I would assume planets in this range to be liquid/gas in their makeup.

In the range from 1,000 M down to 100 M, we find ourselves faced with a surprising situation. There are no known bodies in the Solar System in that mass range. None!

Is that just coincidence, or is there significance to it?

Could it be that the value of 100 M (or something in the neighborhood) is a critical mass?

Could it be, for instance, that if any object condenses into a compact body and has a mass less than 100 M it lacks a gravitational field sufficiently intense to collect and retain the hydrogen that makes up the large majority of the general cosmic cloud out of which stars and planets formed? In that case the object would have to remain small and less than 100 M in mass, since it must be made out of nonhydrogen material to begin with and there isn't enough of that to do much with.

On the other hand, if a compact object happens to be more massive than 100 M, it may possess a gravitational field sufficiently intense to pick up some quantities of hydrogen from the cosmic cloud. The more it picks up, the greater its mass, the more intense its gravitational field, and the more easily it can pick up still more hydrogen. At masses over 100 M, in other words, you would get a "snowball effect" and end up with a body of more than 1,000 M.

It could be for that reason, then, that there are objects less than 100 M and objects more than 1,000 M and nothing in between.

Next let us move to the other end of the scale, that of objects less than 1 M. If we consider known bodies that have masses smaller than that of the Moon, we can list most of the satellites of the Solar System, some hundreds of thousands of asteroids, and uncounted numbers of meteoriods and micrometeoroids.

What they all have in common is that they are solid. Their gravitational fields are far too weak to hold to their surfaces any molecules that, at the prevailing temperatures, are gaseous or liquid. The only materials that can make up such small bodies are metallic or rocky substances made up of atoms held together by interatomic electromagnetic interactions which are enormously stronger than the gravitational interactions such small bodies can produce.

If the body is cold enough it may also be made up

43

of solid substances which at Earthly temperatures are usually thought of as liquid or gas. Such solid volatiles are called "ices."

The only exceptions to this rule of solidity among the minor objects are the comets. Comets formed originally in regions far beyond the planetary orbits, where solar radiation is small enough to be ignored and where the temperature is probably not far above the general level of the background radiation of the Universe—which is only three degrees above absolute zero, or —270° C.

Under those circumstances, everything but helium is a solid and the comets are made up of rocky gravel interspersed with ices, with, in some cases, a rocky core at the center. As long as the rock/ice comets remain in their trans-Plutonian orbits, they are permanent solid bodies, as permanent as the asteroids ringing the Sun between the orbits of Mars and Jupiter.

When, however, gravitational perturbations from the distant stars or the inner planets cause particular comets to take up elongated orbits that carry them into the inner Solar System and relatively near the Sun, the ices evaporate. The object is then solid/gas (or solid/dust/gas).

The comet's gravitational field, being virtually nil, cannot hold any dust or gas that forms and it would quickly diffuse into space and spread through the cometary orbit. As a matter of fact, it is swept away by the solar wind. At each pass near the Sun, more of its substance is swept away and, in the twinkling of an eye on the geologic time scale, it is reduced to the rocky core if it has one, or to nothing if it hasn't.

We might, then, say that any *permanent* object of mass less than 1 M is solid.

Let's move back to the ranges we have skipped. In the range from 100 M to 10 M, we know exactly two bodies, both planets, and they are:

Earth	81.6 M
Venus	69.1 M

Earth and Venus are largely solid, like the bodies of less than 1 M mass, but they have gravitational fields intense enough to retain a gaseous envelope, thin by comparison with the atmospheres of the larger bodies, but significantly thick just the same.

Venus is so hot that none of the major components of its volatile matter can exist as a liquid. It is a solid/gas object.

Earth is cool enough to have water in the liquid state and in large quantity. It is a solid/liquid/gas body. We could argue that life as we know it can only form on a solid/liquid/gas body, though that may just mark our own parochial feelings as to what is right and proper.

In the range from 10 M to 1 M there exist nine known bodies, three planets and six satellites:

Mars	9.0 M
Pluto	9.0 M
Mercury	4.5 M
Ganymede	2.0 M
Titan	1.6 M
Triton	1.5 M
Callisto	1.4 M
Io	1.2 M
Moon	1.0 M

The distinction between planet and satellite is a rather arbitrary one and it seems a shame to lump a large body like Ganymede with an insignificant one like Deimos. I would suggest that the nine bodies in this range be called "subplanets."

The most massive of the subplanets, Mars, does possess an atmosphere, but a thin one. The Martian atmosphere is only a hundredth as dense as that of Earth and only a ten-thousandth as dense as that of Venus. And yet the Martian atmosphere is thick enough to produce mists on occasion, to support dust storms, to provide protection against meteorites—so it deserves the name. Mars is solid/gas.

The least massive of the subplanets, the Moon, is

usually considered to have no atmosphere. Actually, space in the immediate neighborhood of its surface has a density of individual atoms higher than that in regions of outer space far from any large body, so that it might be said to have a "trace atmosphere," one about a trillionth as dense as that of the Earth. This would produce no noticeable effects of the type we generally associate with atmospheres, so for our purposes we will ignore it and think of the Moon as simply solid.

What about the bodies in between? Where is the boundary line between atmosphere and no atmosphere?

The boundary line is hard to draw because in the subplanet range it depends not only on the mass of the body but on its temperature. The higher the temperature, the more rapidly the atoms and molecules of gases move and the more readily they escape into outer space. A given body in the subplanetary range might retain an atmosphere if it were far from the Sun and not do so if it were near the Sun but otherwise unchanged. —So let's consider this matter in greater detail.

In the inner Solar System the only substances likely to form a substantial atmosphere are water, carbon dioxide, and nitrogen.* In the outer Solar System, the only substances likely to form substantial atmospheres on subplanets are water, ammonia, and methane.

Of the subplanets, only Mercury, the Moon, and Mars are in the inner Solar System. Mercury is not much hotter than Venus, which has a thick atmosphere. Mercury, however, has only one-fifteenth the mass of Venus and it lacks the gravitational power to turn the trick. As a cold body it would certainly retain an atmosphere of some sort, but near to the Sun as it is, it cannot. It has at best only a trace atmosphere.

Mars, which has twice the mass of Mercury and is much colder has no trouble retaining an atmosphere. At its low temperature, water is frozen into an ice, so

_____

* I don't mention oxygen because that is not likely to exist in the atmosphere of any world that doesn't bear life. On Earth, it is the product of plant photosynthesis.

its atmosphere contains only carbon dioxide and (probably) nitrogen.

Let us move outward, then, and consider the six subplanets of the outer Solar System. No fewer than three of them, Ganymede, Callisto, and Io, are satellites of Jupiter. (There is a fourth sizable satellite of Jupiter, Europa, but it is less massive than the other three and, with a mass of 0.6 M, falls below the subplanetary range and, by my admittedly arbitrary standards, must be classified as a "minor object.")

Jupiter is 5.2 times as far from the Sun as Earth is, and the temperatures in its satellite system are low enough to freeze not only water (which freezes at 0° C) but ammonia as well (which freezes at —33° C).

As a matter of fact, Ganymede and Callisto have densities only half that of the Moon and only a third that of the Earth. This can be so only if something less dense than rock makes up a sizable portion of their volumes. It may be then that these subplanets are made up largely of ices (water and ammonia).

As for Io (and Europa, too), its density is about that of the Moon, so that it may be largely rock. In its case, though, the surface, at least, is probably covered by a frost of ices.

With water and ammonia eliminated, that leaves only methane as an atmospheric component. Methane doesn't liquefy till a temperature of —162° C is reached and doesn't freeze till one of —182° C is reached. At the temperatures of Jupiter's satellites, then, it is still a gas; and, at that temperature, the gravitational fields of those satellites is still insufficient to hold it. The Jovian satellites, therefore, have no more than trace atmospheres.

(The trace atmosphere of Io has definitely been detected. It is about a billionth as dense as Earth's, but a thousand times as dense as that of the Moon. Oddly enough, it contains sodium—but trace atmospheres can have all sorts of odd components. Substantial ones have to be more serious.)

*   *   *

There are three subplanets in the regions beyond Jupiter. These three are, in order of increasing distance, Titan (the largest satellite of Saturn), Triton (the largest satellite of Neptune), and Pluto (the farthest known object in the Solar System other than comets).

Pluto and Triton are both so far from the Sun that their temperatures are low enough to freeze even methane. The only substances that will remain gaseous at Tritonian and Plutonian temperatures are hydrogen, helium, and neon, and these are so light that even at such low temperatures the gravitational fields of the subplanets are not likely to hold more than traces.

Both bodies are so distant (at their closest each is more than four billion kilometers away) that we are not likely to get direct evidence concerning this for quite a while.

Which leaves us only with Titan—

Titan is the second most massive satellite in the Solar System, second only to Ganymede, and that is a useful property to have if we are looking for an atmosphere. Titan's temperature is about —150° C, fifteen degrees lower than that of Ganymede and the other Jovian satellites. At Titan's temperature, methane is still gaseous, but it is pretty close to its liquefaction point and its molecules are sluggish indeed.

At Titan's temperature, methane could freeze in the form of a loose compound with water, then be released by what internal heat Titan might have, and then be held on to by Titan's gravitational field. The combination of Titan's mass and low temperature would do the trick.

In 1944, the Dutch-American astronomer Gerard Peter Kuiper detected an atmosphere about Titan and found it to consist of methane. Although nowadays we can work out why this is so in hindsight, it was at the time of discovery a titanic surprise (aha!). What's more, the atmosphere is a substantial one, very likely denser than that of Mars.

Titan is the only satellite in the Solar System known to have a real atmosphere, and the only body in the

Solar System to have an atmosphere that is primarily methane.† Methane has been detected in the atmospheres of Jupiter, Saturn, Uranus, and Neptune, but there it exists as a minor component of atmospheres that are primarily hydrogen.

Methane (CH_4) is a carbon compound, and carbon is a unique element.‡ Methane molecules (unlike those of water and ammonia) can break under the lash of solar radiation and recombine into larger molecules. Thus, Pioneer 10 has located in Jupiter's atmosphere not only methane itself, but ethane (C_2H_6), ethylene (C_2H_4), and acetylene (C_2H_2). Undoubtedly, more complicated molecules, with more carbon atoms, exist also, but in successively lower concentrations that make them more difficult to detect. It could be broken fragments of such more complicated carbon compounds that produce the brown and yellow bands on Jupiter and that account for the orange color of Saturn's equatorial regions.

In the giant-planetry atmospheres, however, methane molecules encounter each other relatively rarely, since the superabundant hydrogen molecules get in the way. On Titan, where the atmosphere is almost entirely methane, the reactions may take place more easily. To be sure, the radiation of the distant Sun is feeble, so that the reactions take place more slowly than they would on Earth, for instance, but the reactions, however slow they might be, have had almost five billion years of time in which to proceed.

It may, therefore, turn out that Titan's atmosphere has, as minor constituents, a very complicated mix of organic gases that is responsible for the satellite's orange color. Titan may, indeed, possess an orange cloud cover

‡ See "The One and Only" in *The Tragedy of the Moon* (Doubleday, 1973).

that completely hides its surface. As for that hidden

† There is also hydrogen in Titan's atmosphere, it has recently been found, and since Titan's gravitational field can't hold hydrogen, an interesting theory has been worked out to account for its presence, but that's another story.

surface, it may be covered with a hydrocarbon tar or sludge. Or there may be an ocean of hydrocarbons dissolved in methane (the solution liquefying at a higher temperature than methane itself would), so that Titan might be covered by a petroleum sea.

And might these organic compounds, based on a world that, like Earth, might be solid/liquid/gas, work up into very complex and versatile compounds of a type we know nothing of because they would be too fragile to exist at Earth temperatures? Can there be a cold methane-based life on Titan, to supply us, someday with *another* titanic surprise? Some day, we may find out.

ADDENDUM. You might as well have a summary of my classification of astronomical bodies in this article:

Class of object	Physical state	Typical example
stars	gas	Sun
100,000M		
substars	liquid	Jupiter
10,000M		
giant planets	liquid/gas	Saturn
1,000M		
(empty)		
100M		
planets	solid/liquid/gas	Earth
	solid/gas	Venus
10M		
subplanets	solid/liquid/gas	Titan
	solid/gas	Mars
	solid	Ganymede

Class of object	Physical state	Typical example
1M		
	solid/gas	Halley's Comet
minor objects		
	solid	Europa

4

The Wrong Turning

The other day it was necessary for me, in the course of an article I was writing, to determine how rapidly the innermost edge of Saturn's ring revolved about Saturn.

Since I am a reasonably lazy person, my first thought was to look it up, and I began to go through my reference books, ones on which I had counted quite confidently failed me, I grew annoyed and went through everything. Nothing helped. In many different places I found what the period of revolution of ring particles in Cassini's division would be if there were any there, but nowhere did I find the period of the innermost edge.

I was baffled and for a moment I thought I might revise my article in such a way as no longer to need that piece of information, but that seemed cowardly. I decided to look over the list of Saturn's satellites, their distances, and their periods of revolution and see if I could work out something that would help me with the rings.

I got to work and in five minutes, I had rediscovered Kepler's Third Law.

This plunged me into an embarrassed despair, for you must understand that the prime prerequisite for rediscovering Kepler's Third Law is forgetting it first, and for me to forget it requires a brain of no ordinary stupidity since I had actually written articles dealing with Kepler's Third Law.*

For a while I was too upset to continue working, but

* See "Harmony in Heaven" in *From Earth to Heaven* (Doubleday, 1966).

what is the use in being intelligent if you can't think up a good specious argument to prove the existence of that intelligence against the clearest evidence to the contrary.† I argued this way: An unintelligent fellow wouldn't know that Kepler's Third Law existed. An intelligent fellow would know that Kepler's Third Law existed and would remember it. A superintelligent fellow (aha!) would know that Kepler's Third Law existed but could freely forget it because he could always rediscover it.

This bit of nonsense so heartened me that I not only calculated the period of revolution of the innermost edge of Saturn's rings and went back to work, but I even began to consider how I could use the knowledge to write a different article for my Gentle Readers. And here it is—

Nobody was around to witness the Solar System forming, but a very reasonable conjecture is that it was originally a cloud of dust and gas which gradually coalesced under the influence of its overall gravitational field. The gravitational field intensified steadily as the material condensed, thus hastening further condensation.

Presumably, the condensation produced our present Sun, but it isn't likely that it did so in one smooth process. There must have been subcondensations formed so that there was a stage during the formation of the Solar System when there were innumerable chunks of icy or rocky material formed—colliding, gouging, breaking up, recoalescing, and so on, and most of it gradually ending up in the central body.

It is also reasonable to suppose that as the original cloud of dust and gas coalesced into the centrally situated Sun, which is much less voluminous than the original cloud, the rate of rotation increased. The reason for that is that the angular momentum of a closed system must be conserved. The angular momentum de-

† See Chapter 15.

pends not only on the rate of rotation, but on the distance of the turning object from the center. If that average decreases with condensation, the rate of rotation must increase to make up for it.

As the coalescing Sun turns faster and faster, the centrifugal effect that tends to throw outward from the center grows more marked, particularly where the rate of turn is greatest—at the equatorial region of the coalescing body. As the Sun progresses towards its formation, it becomes an ellipsoid, with matter jutting far out from the equatorial regions in the form of a thinning sheet.‡

The matter within this equatorial sheet can coalesce to form conglomerations of matter smaller than the central body—conglomerations that will continue to move about the Sun, caught in the grip of that body's strong gravitational field and themselves separated by great enough distances from the Sun and from each other to be safe from collisions or near-collisions that would alter their orbits drastically.

And thus we end with planets circling the Sun.

If this is indeed the way in which planets are formed, then we see that they possess certain properties. For instance, they are formed out of that small fraction of the original cloud that represents the equatorial bulge, so planets must be much smaller than the central Sun.

Then, too, the original cloud, including the equatorial bulge, was all turning in one piece, so to speak, so that you would expect the planets to be revolving about the Sun in the same direction that the Sun is rotating about its axis ("direct motion"). Furthermore, the plane of the planetary orbit should be in the plane of the Sun's equator (the "inclination" should be equal to zero) and the planet should be moving a more or less circular orbit (the "eccentricity" should be equal to zero).

‡ Most of the angular momentum of the Solar System ended up in the material formed out of that thinning sheet, which bothered astronomers for a long time. That is something I may deal with in another article some day.

All this is true of the planets. All together, they have a mass about $\frac{1}{750}$ that of the Sun. They each revolve about the Sun in the same direction that the Sun rotates about its axis. They all move in orbits that are nearly circular, and the planes of all the orbits are quite close to that of the solar equator.

The fact that all these things are true can't be coincidence. If the planets were formed with no reference to the Sun at all, they could revolve about the Sun in any plane and with any degree of eccentricity. There is no compelling reason in celestial mechanics why they should not. Comets circle the Sun in any plane and with any eccentricity.

Yet the fact is that each planet moves in direct fashion and with very low eccentricity and inclination. This means that some kind of constraint is involved, something that prevents planets from having high eccentricities and inclinations. It is precisely through an attempt to imagine what the constraint might be that astronomers have thought up this notion of a condensing cloud with an equatorial bulge. It explains the planetary design of the Solar System.

As for the comets, they are the remnants of the original dust cloud that formed the Sun and the planets, remnants so far out that they didn't participate in the condensation. There were sub-condensations into small, icy comets that are now distributed about the Sun in a large hollow sphere, and they are not subjected to the usual planetary constraints for that reason.

The fact that the planets have orbits that are not *exactly* circular and are not *exactly* in the Sun's equatorial plane is not too surprising. They were formed by the slow ingathering of chunks of matter. In general, those chunks arrived from every direction so that the jarring effect canceled out. The last few big ones might, through sheer chance, have been asymmetrically distributed and the nearly fully formed planet might have been given a final knock or knocks that upset its perfect zeroes somewhat.

Naturally, the smaller the planet, the more those last few collisions would have affected it, and it is not surprising that the largest eccentricities and inclinations are found in the smallest planets, Pluto, Mars, Mercury.

If this is truly the way in which the planets of the Solar System were formed, then there might be some marks of those last collisions. Where atmospheres are involved, erosion might wipe them away, and might in any case hide them from our eye. Where the atmosphere is thin, or virtually absent, the last marks are preserved and visible in the form of collision craters. One hemisphere of Mars is rich in them, and Mercury is finely stippled all over with craters.

What applies to the planets should also apply to the satellites. The forming planet should itself have an equatorial bulge, and still smaller bodies should form in the equatorial plane, revolving in direct fashion with nearly zero eccentricity.

Consider Jupiter, for instance. Jupiter has five satellites that revolve in nearly circular orbits in direct fashion in very nearly the planet's equatorial plane. This can't be coincidence; the constraints are there.

In addition to these five satellites, however, Jupiter also has eight other satellites that do not obey the rules. (The eighth was discovered as recently as September 14, 1974.) What of these eight? Do they destroy the entire theory?

No, they don't. These outer eight are very small compared to the others, and very distant from the planet. Even the largest of the outer eight is smaller than the smallest of the inner five. The nearest of the outer eight is over six times as far from Jupiter as the farthest of the inner five. The outer eight are therefore related to Jupiter rather as the comets are related to the Sun; they are not part of the general condensation-equatorial-plane beginning. They are considered to be captured asteroids and as such don't have the usual constraints but can orbit with large eccentricities and inclinations.

The eccentricities of the outer eight range from a

moderate 0.08 (quite small enough to be respectable) to a large 0.38. (Maximum eccentricity is 1.0.) The inclinations vary from 28 degrees to 163 degrees. (Maximum inclination is 180 degrees.) Any inclination between 90 and 180 degrees indicates that the satellite is revolving in the wrong direction and is moving in "retrograde" fashion. Four of Jupiter's outer eight satellites, the four outermost in fact, have retrograde orbits.

The fact that the four outermost have retrograde orbits is a point in favor of the captured-asteroid theory, since it can be shown that the capture of an asteroid into a retrograde orbit is easier than its capture into a direct orbit.

In addition to the outer Jovian satellites, the outermost satellite of Saturn and the outer satellite of Neptune seem to have the characteristics of captured asteroids. The outermost satellite of Saturn is about 3.6 times as far from Saturn as is the next farthest satellite; it has an eccentricity of 0.16 and moves in a retrograde orbit. The average distance of the outer satellite of Neptune is nearly sixteen times that of the inner; it has an inclination of 27.7 degrees (not enough to make it retrograde, but quite high) and an eccentricity of 0.75, higher than that of any other object in the Solar System except for comets.

Astronomers feel quite safe, therefore, in supposing that these ten satellites were not formed out of the same condensing cloud that formed the central planet they circle. That still leaves twenty-three "true satellites" that may have been so formed.

Listing them by planets going out from the Sun, and satellites going out from each planet, the twenty-three are:

Earth, one—the Moon
Mars, two—Phobos, Deimos
Jupiter, five—Amalthea, Io, Europa, Ganymede, Callisto

Saturn, nine—Janus, Mimas, Enceladus, Tethys,
Dione, Rhea, Titan, Hyperion, Iapetus (and also
the rings, of course)
Uranus, five—Miranda, Umbriel, Ariel, Titania,
Oberon
Neptune, one—Triton

Let's consider how these twenty-three satellites fit the
"condensing cloud with the equatorial bulge" hypothesis
as far as their orbital characteristics are concerned. We
can start by considering the distance of the various satellites from their primaries (the planets they circle)—but
not in miles or kilometers. After all, a large planet forms
out of a large coalescing cloud with a large equatorial
bulge and would therefore be expected to have satellites
farther from itself than a small one would. Therefore,
let's measure satellite distance as multiples of the radius
of the primary. This is done in Table 1.

One thing we can point out at once is that, of the ten
satellites considered to be captured satellites, the one
with the smallest distance in terms of primary radii is
Nereid, whose average distance from Neptune is equal
to 130 Neptune radii, a value more than twice as high
as the largest value for any satellite listed in Table 1.
The distance values for the other nine range up to 332
for the outermost of Jupiter's satellites. In terms of distance alone, then, we seem justified in omitting these
ten.

Let's try listing the planets by some logical method
other than sheer distance. As a planet increases in radius, it also increases in mass even more rapidly (barring
an enormous drop in density). It may be that mass
is more important than mere radius, since it is mass that
is the source of the planet's gravitational field, and it is
the strength of the gravitational grip that keeps the
bulge in line and produces satellites that do not deviate
much, if at all, from the equatorial plane and from or-
bit-circularity. After all, two satellites may each be at

58

Table 1—Satellite Distance

Satellite	Distance (primary radii)
Rings (inner edge)	1.24
Rings (outer edge)	2.28
Amalthea	2.54
Janus	2.64
Phobos	2.71
Mimas	3.10
Enceladus	3.99
Tethys	4.94
Miranda	5.44
Io	5.91
Dione	6.32
Deimos	6.95
Ariel	8.41
Rhea	8.83
Europa	9.40
Umbriel	11.7
Triton	13.4
Ganymede	15.0
Titania	19.9
Titan	20.5
Hyperion	24.8
Oberon	25.7
Callisto	26.4
Iapetus	59.6
Moon	60.3

a distance of five times the radius of its primary, but the more massive planet will exert the greater gravitational effect at that distance.

The most marked effect of the planetary gravitational field involves the speed at which a satellite moves in its orbit. Therefore, let us list the satellites again, this time in the order of orbital speed and see if that makes any

Table 2—Satellite Orbital Speed

Satellite	Orbital speed (kilometers/second)
Amalthea	13.15
Rings (inner edge)	10.74
Io	8.66
Rings (outer edge)	8.37
Janus	7.75
Mimas	7.16
Europa	6.84
Enceladus	6.33
Tethys	5.66
Ganymede	5.44
Dione	4.98
Rhea	4.23
Callisto	4.10
Miranda	3.19
Titan	2.78
Ariel	2.75
Hyperion	2.53
Umbriel	2.34
Triton	2.20
Titania	1.89
Iapetus	1.64
Oberon	1.59
Phobos	1.04
Deimos	0.68
Moon	0.51

marked difference in the distance characteristic. This is done in Table 2.

As you see, the chief difference between Tables 1 and 2 is that the satellites of massive Jupiter move further toward the head of the list while those of small Mars drop toward the bottom. In Table 1, Iapetus and the

Moon are so much farther away than the others that we might have doubted their status, but in Table 2, the spread evens out, although the Moon is still at the bottom of the list.

Next, let's consider the eccentricity of each satellite (its departure from orbital circularity) and the inclination of the orbit to the equatorial plane of its primary. If the theory of satellite formation from the equatorial bulge of the coalescing planet is correct, then both

Table 3—Satellite Eccentricity and Inclination

Satellite	Eccentricity	Inclination (°)
Amalthea	0.00	0.1
Rings (inner)	0.00	0.0
Io	0.00	0.1
Rings (outer)	0.00	0.0
Janus	0.00	0.0
Mimas	0.02	1.5
Europa	0.00	0.1
Enceladus	0.00	0.0
Tethys	0.00	1.1
Ganymede	0.00	0.3
Dione	0.00	0.0
Rhea	0.00	0.3
Callisto	0.01	0.2
Miranda	0.00	0.0
Titan	0.03	0.3
Ariel	0.01	0.0
Hyperion	0.10	0.6
Umbriel	0.01	0.0
Triton	0.00	27.7
Titania	0.02	0.0
Iapetus	0.03	14.7
Oberon	0.01	0.0
Phobos	0.02	1.1
Deimos	0.00	1.8
Moon	0.06	23.5

values should be, ideally, zero. The actual values, given to two decimal places in the case of eccentricity, and one in the case of inclination, are given in Table 3 (with the satellites listed in the order given in Table 2).

Actually, as you can see, most of the satellites come quite close to the ideal—close enough to make it quite certain that it couldn't happen by coincidence in so many cases. Only the equatorial bulge (or something equally good that no astronomer has yet happened to think of) could account for it.

As a matter of fact, the satellites fit the hypothesis even better than the planets do.

The nine planets have, in some cases, moderately high orbital eccentricities. That of Pluto is 0.25 and of Mercury 0.21. The average eccentricity for the nine planets is 0.08. That is not high, but the average eccentricity for the twenty-five satellites (and rings) listed in Table 3 is only 0.016. The satellite with the most lopsided orbit is Hyperion, which has an eccentricity of 0.10, and its orbit is only slightly more eccentric than that of Mars (0.093) and has nothing like the eccentricity of Mercury and Pluto.

Inclination is not quite so clear-cut. The inclinations of the planetary orbits can deviate by several degrees from the ideal. Earth's orbit is inclined seven degrees to the Sun's equatorial plane, and if Earth's orbit is taken as the standard, the inclinations of the other planets deviate by a few degrees, the figure being highest for Pluto, which has an inclination to Earth's orbit of seventeen degrees.

In comparison with this, twenty-two of the objects listed in Table 3 have inclinations to the equatorial plane of their primary of less than two degrees and ten of them have an inclination of less than a tenth of a degree. There is no way of getting round that without calling on the equatorial bulge theory.

And yet some satellites do represent a puzzle. Let us concentrate on those satellites which have an eccentricity of higher than 0.08 (the planetary average) or an inclination of greater than two degrees, or both. These are listed in Table 4, and there are only four of them.

Hyperion is not very impressive in its irregularity, as I have already said. Its eccentricity is only marginally high and its inclination is quite satisfactorily low. We can let it go.

The Moon is a special case, which I discussed in an

Table 4—The Odd Satellites

Satellite	Eccentricity	Inclination (°)
Hyperion	0.10	0.6
Triton	0.00	27.7
Iapetus	0.03	14.7
Moon	0.06	23.5

earlier article* and I won't linger on the matter here. It may be, after all, a captured body, which would account for its high inclination and its marginally high eccentricity. Then, too, Earth and Moon affect each other tidally to an unprecedented amount since the two bodies are far more nearly equal in size than is any other satellite-primary combination in the Solar System (or any planet-Sun combination either). The tidal effects may have modified whatever the original orbit may have been and produced the present odd situation.

As for Iapetus, that is a rather unusual satellite. When it is to the west of Saturn, it is six times as bright as when it is to the east of Saturn If it turns only one face to Saturn, as the Moon does to Earth (something that seems a reasonable assumption) then we are seeing one hemisphere when it is to the west and the other hemisphere to the east.

There must therefore be a profound asymmetry to the satellite (which is of moderate size—perhaps 1,750 kilometers in diameter, half that of the Moon). Further-

* "Just Mooning Around" in Of Time and Space and Other Things (Doubleday, 1965).

more it must be that the asymmetry is distributed in such a way that the hemisphere we see when Iapetus is on one side of Saturn is quite different from the other; that we see each different hemisphere head on, or nearly so.

Whatever the asymmetry is, it must be something that is capable of making one hemisphere icy and reflective and the other rocky and nonreflective. Perhaps the asymmetry is the result of an unusually severe final blow in the coalescence of the satellite so that Iapetus is a kind of double world, the lesser half forming a large lump on the larger, with one of the two icy and the other rocky. And it may be this final blow that knocked Iapetus well out of the equatorial plane. (I have not seen this notion advanced anywhere, so if it is a poor theory the blame is mine.)

That leaves us with Triton, which is more highly inclined than are either the Moon or Iapetus, but which has a virtually circular orbit as opposed to the marginal eccentricities of the other two. This combination of high inclination (the highest for any satellite that is not clearly a captured one) and very low eccentricity is a curious one, so let's look at Triton a little more closely.

Triton was discovered in 1846 only a month after Neptune itself was discovered, and small wonder, for it is one of the large satellites. It has a diameter of 3,700 kilometers, a trifle larger than the Moon, and anyone looking at Neptune with a good telescope would see Triton without much trouble.

When Triton was first discovered, it was observed to be revolving in the retrograde direction. It was assumed that this was because Neptune itself was rotating on its axis in a retrograde direction.

You see, if the condensing-cloud-with-equatorial-bulge theory is right, then, ideally, every planet ought to rotate in the direct fashion, with its axis at right angles to its plane of revolution. However, for some reason, planetary rotation tends to deviate quite a bit from the ideal. The axis of Jupiter, to be sure, tips only 3.1 degrees from this perpendicular ideal, but in the case of Mars and Saturn the axial tipping is 25.2 degrees and 26.7 degrees respectively, and that of Uranus is 98.0 degrees.

Yet the satellite system tips with it. The satellites of Mars and Saturn remain in the equatorial plane of their primary, and so do the satellites of Uranus. Uranus seems to be rotating on its side, so to speak, so that when it is properly oriented in its orbit it seems to have an east pole and a west pole relative to ourselves. The Uranian satellites go along with this and seem to revolve in an undown direction relative to ourselves whereas all others seem to revolve in a right-left direction more or less.

Whatever it was that tipped the planet, the effect must have taken place while the planetary cloud was still coalescing, so that the equatorial bulge tipped with it.

It seemed then that Neptune's cloud had tipped extremely—over 150 degrees so that it was standing virtually on its head and therefore rotating in the wrong direction, with Triton faithfully following that wrong direction. (Venus stands on its head, but it has no satellite, so we can't have a closer example of a satellite standing on its head along with its primary.)

But then, in 1928, Neptune was studied carefully by means of a spectroscope, which would show which side of it was coming forward toward Earth and which side was receding—and it turned out that Neptune was rotating about its axis in direct fashion. Its axial tilt was only 29 degrees.

This meant Triton was making a wrong turn without excuse. It was standing on its head by itself. Its inclination was not 27.7 degrees at all but 152.3 degrees to indicate its retrograde revolution.

Triton's inclination is roughly that of the four outermost satellites of Jupiter and of Phoebe, the outermost satellite of Saturn. Those five outer satellites of the two giant planets are, however, generally agreed upon as captured satellites. Does that mean Triton is a captured satellite as well?

But if Triton is a captured satellite, how is it possible

that by some absolutely unbelievable coincidence, it just happened to settle down into an almost perfectly circular orbit? None of the ten captured satellites have orbits that are even near to circular. The ten captured satellites have an average orbital eccentricity of 0.25 and the least eccentric of the ten has an eccentricity of 0.08. Can Triton be reasonably expected to have an eccentricity of 0.00 if it is a captured satellite?

Yet if Triton was formed in the equatorial bulge so as to have a virtually zero eccentricity, how is it possible for it to be making the wrong turning—moving against the current, so to speak, of the materials forming Neptune?

Triton, it would seem, is the most puzzling of the satellites—more puzzling than the Moon, even.

5

The Bridge of the Gods

On June 6, 1974, my wife, Janet, and I were in the Forest of Dean, in southwestern England near the Welsh border. It was a day of showers interspersed with sunshine, and in the late afternoon Janet and I took a walk among the immemorial beeches.

A sprinkle of rain sent us under one of those beeches but the Sun was out and a rainbow appeared in the sky. Not one rainbow, either, but *two*. For the only time in my life I saw both the primary and secondary bows, separated, as they should be, by about twenty times the diameter of the full Moon. Between them, the sky was distinctly dark, so that, in effect, we saw a broad band of darkness crossing the eastern sky in a perfect circular arc, bounded on either side by a rainbow, with the red side of each bordering the darkness and the violet side fading into the blue.

It lasted several minutes and we watched in perfect silence. I am not a visual person, but that penetrated—and deeply.

Nine days later, on June 15, 1974, I visited Westminster Abbey in London and stood beside Isaac Newton's grave (I refused to step *on* it.) From where I stood, I could also see the graves of Michael Faraday, Ernest Rutherford, James Clerk-Maxwell, and Charles Darwin; all told, five of the ten men whom I once listed* as the greatest scientists of all time. It penetrated as deeply as the double rainbow.

---

* See "The Isaac Winners" in *Adding a Dimension* (Doubleday, 1964).

I couldn't help thinking of the connection between the rainbow and Newton and decided at once to do an article on the subject when the occasion lent itself to the task—and here it is.

Suppose we begin with light itself. In ancient times, those we know of who speculated on the matter thought of light as preeminently the property of the heavenly bodies and, in particular, of the Sun. This heavenly light was not to be confused with Earthly imitations such as the fire of burning wood or of a burning candle. Earthly light was imperfect. It flickered and died; or it could be fed and renewed. The heavenly light of the Sun was eternal and steady.

In Milton's *Paradise Lost* one gets the definite impression that the Sun is simply a container into which God has placed light. The light contained in the Sun is forever undiminished, and by the light of that light (if you see what I mean) we can see. From that point of view, there is no puzzle in the fact that God created light on the first day and the Sun, Moon, and stars on the fourth. Light is the thing itself, the heavenly bodies merely the containers.

Since sunlight was heavenly-born it would naturally have to be divinely pure, and its purity was best exemplified in the fact that it was perfectly white. Earthly "light," imperfect as it was, could have color. The flames of Earthly fires were distinctly yellowish, sometimes reddish. Where certain chemicals were added, they could be any color.

Color, in fact, was an attribute, it seemed, of Earthly materials only, and when it intruded into light, it seemed invariably a sign of impurity. Light reflected from an opaque colored object, or transmitted through a transparent colored object, took on the color and imperfection of matter, just as clear water coursing over loose silt would grow muddy.

There was only one aspect of color which, to the eyes of the ancients, did not seem to involve the kind of matter they were familiar with, and that was the rainbow. It appeared in the sky as a luminous arc of different colors: red, orange, yellow, green, blue, and violet, in that order, with the red on the outer curve of the arc and the violet on the inner curve.†

The rainbow, high in the sky, insubstantial, evanescent, divorced from any obvious connection with matter, seemed as much an example of divine light as that of the Sun—and yet it was colored. There was no good explanation for that except to suppose that it was another creation of God or of the gods, produced in color for some definite purpose.

In the Bible, for instance, the rainbow was created after the Flood. God explained its purpose to Noah: "And it shall come to pass, when I bring a cloud over the earth, that the bow shall be seen in the cloud: And I will remember my covenant, which is between me and you and every living creature of all flesh; and the waters shall no more become a flood to destroy all flesh" (Genesis 9:14—15).

Presumably, though the Bible doesn't say so, the rainbow is colored so that it can the more easily be seen against the sky, and serve as a clearer reassurance to men trembling before the wrath of God.

The Greeks took a less dramatic view of the rainbow. Since it reached high in the sky and yet seemed to approach the Earth at the either end, it seemed to be a connecting link between Heaven and Earth. It was the bridge of the gods (colored, perhaps, because it was a material object, even though of divine origin) whereby they could come down to Earth and return to Heaven.

In Homer's *Iliad*, the goddess Iris is the messenger of the gods, and comes down from Olympus now and then to run some errand or other. But *iris* is the Greek word

† A seventh color is often added, "indigo." To my eyes, indigo is only a bluish-violet and does not deserve the dignity of a separate color of the rainbow. The presence of an indigo-colored component of the light emitted by a certain ore heated to incandescence revealed a new element, however, which was consequently named "indium."

for "rainbow" (and because that portion of the eye immediately about the pupil comes in different colors, it, too, is called the iris). The genitive form of the word is *iridis,* and when there is a colored, rainbowlike shimmering on matter, as on a soap bubble, it is said to be "iridescent." And because the compounds of a certain new element showed a surprising range of color, the element was named "iridium."

In the Norse myths, the rainbow was "Bifrost" and it was the bridge over which the gods could travel to Earth. Before the last battle, Ragnarok, it was one of the signs of the coming universal destruction that under the weight of the heroes charging from Valhalla, the rainbow bridge broke.

But what about rational explanations? Steps were made toward those, too. In ancient times, the Greek philosopher Aristotle, about 350 B.C., noted a rainbow effect seen through a spray of water—the same colors in the same arrangement and just as insubstantial. Perhaps the rainbow itself, appearing after rain, was produced in similar fashion by water droplets high in the air.

Nor was water the only transparent substance associated with the rainbow. About A.D. 10 or so, the Roman philosopher Seneca wrote of the rainbowlike effect of colors that showed on the broken edge of a piece of glass.

But what is there about light and transparent substances that can produce a rainbow? It is quite obvious that light passing through such substances in ordinary fashion produces no colors. There is, however, a certain peculiarity in the way light behaves when it crosses from one type of transparent substance to another—from air to water, for instance—that might offer a clue.

This peculiar behavior first entered the history of science when Aristotle pointed out what innumerable people must have casually noticed: that a stick placed into a bowl of water seems to be bent sharply at the water surface, almost as though it were broken back into an angle at that point. Aristotle attributed this to the bending of light as it passed from air into water, or from water into air. After all, the stick itself was not really bent, since it could be withdrawn from the water and shown to be as straight as ever—or felt while it was still in the water and experienced as still straight. The bending of light in passing from one medium to another is called "refraction" (from Latin words meaning "breaking back").

Could it be that the rather unusual event of color-formation by water or glass could involve the rather unusual fact of the changing of direction of a beam of light?

The first person actually to suggest this was a Polish monk named Erazm Ciolek, in a book on optics which he wrote in 1269 under the partially Latinized name Erasmus Vitellio.

Merely to say that refraction was responsible for the rainbow is easy. To work out exactly how refraction could result in an arc of the precise curvature and in the precise position in the sky is an altogether more difficult thing to do and it took three and a half centuries after the refraction suggestion was made for someone to dare work it out mathematically.

In 1611, Marco Antonio de Dominis, Archbishop of Spalato (who was imprisoned by the Inquisition toward the end of his life because he was a convert to Anglicanism and argued against Papal supremacy) was the first to try; but managed only a very imperfect job. Unfortunately, ever since Greek times, people had had an inaccurate idea as to the precise manner in which light was refracted— and so did the Archbishop.

It was not until 1621 that refraction was finally understood. In that year, a Dutch mathematician, Willebrord Snell, studied the angle which a beam of light made with the perpendicular to the water surface it was entering, and the different angle it made with the perpendicular once it was within the water. It had been thought for many centuries that as one angle changed, the other angle changed in proportion. Snell showed that it is the sines† of the angles that always bears the same ratio, and this constant ratio is called "the index of refraction."

Once the notion of an index of refraction was known, scientists could trace the path of light through spherical water droplets, allowing for both reflection and refraction, with considerable precision.

This was done by the French philosopher René Descartes in 1637. He used Snell's Law to work out the precise position and curvature of the rainbow. However, he did not give the proper credit to Snell for the law but tried to leave the impression, without actually saying so, that he had worked it out himself.

Snell's Law, however, did not, in itself, properly explain the *colors* of the rainbow.

There seemed only two alternatives. First, it was possible that the color arose, somehow, out of the colorless water (or glass) through which the light passed. Second, it was possible that the color arose, somehow, out of the colorless light as it passed through the water (or glass.)

Both alternatives seemed very unlikely since, in either case, color had to derive from colorlessness, but there was a tendency to choose the first alternative, since it was better to tamper with water and glass than with the holy light of the Sun.

The Sun and its light had so often been touted as a symbol of Deity (not only in Christian times, but in pre-Christian times dating back to the Egyptian pharaoh Ikhnaton in 1360 B.C., and who knows how much farther back to what dim speculations of prehistoric time) that it had come to seem, rather foolishly, that to im-

---

† In these articles I try to explain every concept I use as I come to it, but a line has to be drawn. Sines, and trigonometric functions generally, deserve an entire article to themselves and some day I'll write one. Meanwhile, if you don't know what sines are, it doesn't matter. They play no further part in the present argument.

pute imperfection to the Sun and sunlight was to deny the perfection of God.

Consider what happened to Galileo, for instance. There was a number of reasons why he got into trouble with the Inquisition, the chief of them being that he could never conceal his contempt for those less intelligent than himself, even when they were in a position to do him great harm. But it helped that he gave them weapons with which to attack him, and perhaps the chief of these was his discovery of dark spots on the Sun.

He had noted sunspots first toward the end of 1610, but made his official announcement in 1612 and presented a copy of his book on the subject to Cardinal Maffeo Barberini, who was then a friend of his but who from that time (for various reasons) slowly began to cool toward him and who was Pope Urban VIII and an outright enemy when, twenty years later, Galileo's troubles with the Inquisition reached their climax.

The finding of sunspots (and the reality of that finding was irrelevant) offended those mystics who found the Sun to be a type of God, and some began to preach against him.

One of them was a Dominican friar who made use, very tellingly, of an amazingly apt quotation from the Bible. At the beginning of the Acts of the Apostles, the resurrected Jesus finally ascends to heaven and his Galilean apostles stare steadfastly upward at the point where he disappeared until two angels recall them to their Earthly duties with a reproof that begins with "Ye men of Galilee, why stand ye gazing up into heaven?"

In Latin, the first two words of the quotation are *Viri Galilaei,* and Galileo's family name was Galilei. In 1613 when the Dominican thundered out that phrase and used it as a biblical denunciation of Galileo's attempts to penetrate the mysteries of the heavens, many must have shuddered away from the angel-reproved astronomer. In 1615, Galileo's case was in the hands of the Inquisition and his long ordeal began.

*　　*　　*

Yet sunspots can be explained away. Their presence need not be accepted as a final disruption of Heaven's perfection. If the Sun is only the container of light, it might be imperfect and smudged. The thing contained, however, the heavenly light itself, the first creation of God on the First Day, was another matter altogether. Who would dare deny *its* perfection?

That blasphemy came about in England in 1666, a place and time much safer for the purpose than the Italy of 1612. And the man who carried through the blasphemy was a quite pious twenty-four-year-old named Isaac Newton.

The young Newton was interested in the rainbow effect not for its own sake but in connection with a more practical problem which concerned him but does not, at the moment, concern us.

Newton might have begun by arguing that if a rainbow is formed by the refraction of light by water drops, then it should also be formed in the laboratory, if refraction were carried through properly. Refraction takes place when light passes from air into glass at an oblique angle, but if the glass surface is bounded by two parallel planes (as ordinary window glass is, for instance) then, on emerging from the other surface, the same refraction takes place in reverse. The two refractions cancel and the ray of light passes through unrefracted.

One must therefore use a glass object with surfaces that are not parallel and that refract the light entering the glass and the light leaving the glass in the same direction, so that the two effects add on instead of canceling.

For the purpose, Newton used a triangular prism of glass, which he knew, by Snell's Law, would refract light in the same direction on entering and on leaving, as he wanted it to do. He then darkened a room by covering the windows with shutters and made one little opening in one shutter to allow a single circular beam of light to enter and fall on the white wall opposite. A brilliant circle of white light appeared on the wall, of course.

Newton then placed the prism in the path of the light and the beam was refracted sharply. Its path was bent and the circle of white light was no longer where it had been but now struck the wall in a markedly different position.

What's more, it was no longer a circle but an oblong some five times longer than it was wide. Still more, colors had appeared, the same colors as in the rainbow and in the same order.

Was it possible that this rainbow was just a lucky freak resulting from the size of the hole or the position of the prism? He tried holes of different sizes and found that the artificial rainbow might get brighter or dimmer but the colors remained, and in the same order. They also remained if he had the light pass through the thicker or thinner part of the prism. He even tried the prism outside the window so that the sunlight went through it *before* it went through the hole in the shutter—and the rainbow still appeared.

So far, these experiments, though they had never been conducted with anything like such systematic care, did not introduce anything completely new. After all, rainbow effects had, for centuries, been observed and reported at oblique edges of glass which had been either broken or beveled, and that was essentially what Newton was now observing.

It had always been assumed before, though, that the effects were produced by the glass, and now Newton found himself wondering if that could possibly be so. The fact that changing the position of the glass or the thickness of the glass through which the light passed did not change the rainbow in any essential way made it seem the glass might not be involved; that it was the light itself that might be responsible.

It seemed to Newton that if he held the prism point down and then had the light that had passed through it pass through a second prism oriented in the opposite direction, with the point up, one of two things ought to happen:

1) If it was the glass producing the colors as light was refracted through it, more color would be produced by the glass of the second prism and the colored oblong of light would be still more elongated, and still more deeply colored.

2) If it was refraction alone that produced the colors and if the glass had nothing to do with it, then the second refraction, being opposite in direction, should cancel out the first so that the oblong would be a circle again with all the colors gone.

Newton tried the experiment and the second alternative seemed to be it. The light, passing through two prisms that were identical except for being oppositely oriented, struck the wall where it would have struck if there had been no prisms at all, and struck it as a brilliant circle of pure white light. (If Newton had placed a piece of white cardboard between the prisms he would have seen that the oblong of colors still existed there.)

Newton decided, therefore, that the glass had nothing to do with the color, but served only as a vehicle of refraction. The colors were produced out of the sunlight itself.

Newton had, for the first time in man's history, clearly demonstrated the existence of color apart from matter. The colors he had produced with his prism were not colored this or colored that; they were not even colored air. They were *colored light,* as insubstantial and as immaterial as sunlight itself. Compared to the gross and palpable colored matter with which people had been familiar till then, the colors Newton had produced were a kind of ghost of color. It's not surprising, then, that the word he introduced for the band of colors was the Latin word for ghost—"spectrum."*

Newton went on to allow his beam of refracted light

---

* We still speak of "specters" and "spectral appearances" but the new meaning of the word, signifying a whole stretch of different colors, has taken over and is now a common metaphor. We can speak of "the spectrum of political attitudes" for instance.

to fall on a board with a hole in it so that only the single color of a small portion of the spectrum could pass through. This single-color portion of sunlight he passed through a second prism and found that although it was broadened somewhat, no new colors appeared. He also measured the degree to which each individual color was refracted by the second prism, and found that red was always refracted less than orange, which was refracted less than yellow, and so on.

His final conclusion, then, was that sunlight (and white light generally) is not pure but is a mixture of colors, each of which is much more nearly pure than white light is. No one color by itself can appear white, but all of them together, properly mixed, will do so.

Newton further suggested that each different color has a different index of refraction in glass or in water. When light passes through a glass prism or through water droplets, the differences in index of refraction cause the different colored components of white light to bend each by a different amount and emerge from glass or water separated.

This was the final blow to the ancient/medieval view of the perfection of the heavens. The rainbow, that reminder of God's mercy, that bridge of the gods, was reduced to a giant spectrum high in the air, produced by countless tiny prisms (in the form of water droplets) all combining their effect.

To those who value the vision of the human mind organizing observations into natural law and then using natural law to grasp the workings of what had until then been mysterious, the rainbow has gained added significance and beauty through Newton's discovery, because, to a far greater extent than before, it can be *understood* and truly appreciated. To those of more limited fancy, who prefer mindless staring to understanding, and simple-minded fairy tales of gods crossing bridges to the dancing changes of direction of light in accordance with a system that can be written as an elegant mathematical expression, I suppose it is a loss.

Newton's announcement of his discoveries did not take the world by storm at once. It was so revolutionary, so opposed to what had been taken for granted for many centuries, that many hesitated.

For instance, there was the opposition of Robert Hooke, seven years Newton's senior, and with an important position at the Royal Society, which was the arbiter of science in those days. Hooke had been a sickly youngster. Smallpox had scarred his skin, but he had had to work his way through Oxford waiting on tables, and the scapegoatings and humiliations he had to endure at the hands of the young gentry who were infinitely his inferiors intellectually left deeper marks on him than the smallpox did.

The world was his enemy after that. He was one of the most brilliant scientific thinkers of his time and might easily have ranked a clear second after Newton himself if he had not put so much of his time into a delighted orgy of spiteful disputation.

In particular, he marked down Newton for his prey, out of sheer jealousy of the one man whose intellectual equal he could never be. Hooke used his position in the Royal Society to thwart Newton at every turn. He accused him of stealing his (Hooke's) ideas, and nearly kept Newton's masterpiece, *Principia Mathematica,* in which the laws of motion and of universal gravitation are expounded, from being published, through such an accusation. When the book was published at last, it was not under Royal Society auspices, but at the private expense of Newton's friend Edmund Halley.

Newton, who was a moral coward, incapable of facing opposition openly although he was willing to use his friends for the purpose, and who was given to sniveling self-pity, was cowed and tormented by the raging, spiteful Hooke. At times, Newton would vow he would engage in no more scientific research, and in the end, he was driven into a mental breakdown.

It wasn't till Hooke's death that Newton was willing

to publish his book *Opticks,* in which he finally organized all his optical findings. This book, published in 1704, was in English rather than in Latin as *Principia Mathematica* had been. Some have suggested that this was done deliberately in order to limit the extent to which it would be read outside England and therefore cut down on the controversies that would arise, since Newton, for various reasons, was not entirely a popular figure on the Continent.

Opposition to the notion of white light as a mixture of colors did not disappear altogether even after the appearance of *Opticks.* As late as 1810 a German book entitled *Farbenlehre* ("Color-science") appeared and argued the case for white light being pure and unmixed. Its author was none other than the greatest of all German poets, Johann Wolfgang von Goethe, who, as a matter of fact, had done respectable scientific work.

Goethe was wrong, however, and his book dropped into the oblivion it deserved. It is only remembered now as the last dying wail against Newton's optical revolution.

Yet there is this peculiar point to be made. Newton's optical experiments, as I said earlier, were not carried through solely for the purpose of explaining the rainbow. Newton was far more interested in seeing whether there was any way of correcting a basic defect in the telescopes that, ever since Galileo's time a half-century before, had been used to study the heavens.

Till then, all the telescopes had used lenses that refracted light and that produced images that were fringed with color. Newton's experiments seemed to him to prove that color was inevitably produced by the spectrum-forming process of refraction and that no refracting telescope could possibly avoid these colored fringes.

Newton therefore went on to devise a telescope that made use of mirrors and reflection, thus introducing the reflecting telescope that today dominates the field of optical astronomy.

Yet Newton was wrong when he decided that refracting telescopes could never avoid those colored fringes. You see, in his marvellous optical experiments he had overlooked one small thing.—But that is another story.

6

The Third Liquid

A few nights ago, I had occasion (not entirely of my own volition) to be present at a very posh apartment on Manhattan's East Side where a sit-down dinner for twenty was being given under conditions of elegance to which I am entirely unaccustomed.

I was at one of the three tables and, as a quasi celebrity, I was questioned concerning my work. I answered the inevitable question, "Are you writing anything now?" with the equally inevitable, "Yes, of course." That is, and must be, my answer on any hour of any day of these last thirty years.

The hostess said, "And what are you writing today?"

"Today," I said, "I am writing my monthly column for *American Way*, the in-flight magazine of American Airlines."

"Oh?" said she, politely. "And what does the column deal with?"

"The overall title of the column series is *Change*," I said, "and each article deals with some aspect of the future as I see it."

She all but clapped her hands in glee, and said, "Oh, you foresee the future! You believe in astrology!"

For a moment I was taken aback, and then I said, sternly, "I do *not* believe in astrology."

Everyone at the table (each of whom was a quasi intellectual, in order, I presume, to match my position as quasi celebrity) turned to me in horror and cried out in disbelief, "You don't believe in astrology?"

"No," I said, even more sternly. "I don't."

So I was ignored for the rest of the meal, while the

others competed with each other in an effort to see which one could intellectualize most quasily. It was a grim evening.

You will not blame me, Gentle Reader, if I do not, then, allow this article to deal with the planets, as four of the last five have, in order to avoid even the distant suggestion of astrological thinking. I will, instead, turn to another favorite subject of mine—the chemical elements.

There are 105 elements known today and eleven of them are gases at the temperatures to which we are accustomed in our everyday affairs. Six of them are the noble gases: helium, neon, argon, krypton, xenon, and radon. The other five are the more or less ignoble gases hydrogen, nitrogen, oxygen, fluorine, and chlorine.

Of the remaining ninety-four elements, exactly two are usually listed as liquid.

One of these is the metal mercury, which has been known since ancient times. As the only metal to be liquid (and a particularly dense metal, too), it was an impressive substance to the early chemists, who suspected it to be a key to the transmutation of the elements.

Other liquids were known (water, turpentine, olive oil), but no other liquid element was discovered until 1824, when a French chemist, Antoine Jérôme Balard, discovered a red-brown liquid, more or less by accident, while extracting substances from plants growing in a salt marsh.

Whereas mercury boils at 356.6° C and gives off little vapor at room temperature, the new liquid element boils at 58.8° C, and at the temperature of a warm day (25° C or 77° Fahrenheit) produces a vapor which is very noticeable, since it is reddish in color. A stoppered bottle of clear glass, half full of the liquid, is red all the way up.

The vapor has a strong odor, which is usually described as disagreeable, and so the element was named from the Greek word *bromos,* meaning "stench." The element is "bromine."

Reaction to smells is subjective. I, myself, find the smell of bromine strong and not exactly pleasant, but I do not find it disagreeable and certainly do not consider it a stench. Nevertheless, I do not smell it voluntarily, for bromine is an extremely active element, and its vapor will irritate and damage those parts of the body with which it comes in contact.

Whereas mercury freezes at —38.9° C, bromine freezes at —7.2° C. A cold winter's day in New York will suffice to freeze bromine, but it would take a cold winter's day in North Dakota or Alaska to freeze mercury.

This demonstrates the subjectivity of defining elements as solids, liquids, and gases. If the customary temperature at which we lived were —10° C, we'd feel that mercury was the only liquid element while bromine would seem to us to be an easily melted solid. If the customary temperature were —35° C, there'd be two liquid elements, because mercury would still be liquid and chlorine (to us a gas) would have liquefied. At —45° C there would be no liquid elements, but at lower temperatures still, radon would liquefy, and so on.

In order to make sense, then, of the subject matter of this article, let's define a liquid element as one that is liquid at 25° C, which is a little on the warm side but is a common enough temperature in New York City, for instance.

With that settled, and leaving mercury and bromine to one side, what are the nearest misses to liquidity in the rest of the list of elements?

The near misses might be those of gaseous elements that don't quite liquefy at 25° C, or solid elements that don't quite melt. In the first group, there are no really near misses. Of the gaseous elements, the one with the highest liquefying point is chlorine, which manages to become liquid at —34.6° C, some sixty degrees below

25° C. It has probably never been cold enough in New York (in historic times, anyway) to liquefy chlorine.

In the second group, we can do better. There is gallium, for instance, a solid metal, silvery in appearance when pure, which has its melting point at 29.8° C. This is less than five degrees above 25° C and is equivalent to 85.6° Fahrenheit, so that it would be liquid on a hot day in July in New York—and would, in fact, be liquid at body temperature, and would melt in one's hand.

Yet we can do better than that—but let me begin at the beginning.

It was discovered in ancient times that if plants were burned and the ashes were mixed with water, a substance would be dissolved in the water which could then be recovered when the water was evaporated. The substance was useful in preparing such substances as soap and glass. Since the ash extract was usually evaporated in a large pot, the resultant substance was called "potash" in straightforward English.

The Arabs, who were the chemists par excellence in the Middle Ages, called it *al-quili* in straightforward Arabic, since that meant "the ashes." European chemists borrowed many of the Arabic terms when they translated Arabic books on chemistry, so that potash became an example of an "alkali."

Some seashore plants produced an ash that yielded something similar to potash but not identical with it, something that was even better for making soap and glass. The Arabs called this ash *natron,* adopting that from an earlier Greek term, *nitron.* The Arabs must have used natron as a headache remedy (for it neutralized excess acid in the stomach without too badly damaging it). Since their word for a splitting headache was *suda,* that came to be applied to the substance and the word became "soda" in Europe.

Both potash and soda are carbonates of certain metals which were unknown prior to 1800 because those metals were so active and clung to other elements so tightly that they could not be isolated. In 1807, however, the English chemist Humphry Davy used an electric current to split those elements away from their compounds and obtained them in free form for the first time.

The metallic element of potash, he called "potassium," using the by-then conventional suffix for a metallic element. The metallic element of soda, he called "sodium." The Germans, however, named the element of "alkali" (the alternate name for potash) "kalium," and the element of "natron" (the alternate name for soda) "natrium." What's more, the German influence in chemistry in the early nineteenth century was such that the international chemical symbols for those elements are taken from the German names, not the English. The symbol for potassium is "K" and for sodium is "Na" and that's that.

In English, potash is now more properly known as "potassium carbonate," while soda is "sodium carbonate." Both are considered alkalis and the characteristic properties of the two substances (the ability to neutralize acids, for instance) are described as "alkaline." As for potassium and sodium, they are examples of "alkali metals," and they resemble each other closely.

Once chemists worked out the electronic structure of the atoms of the various elements, it could be seen why potassium and sodium so closely resembled each other. Each sodium atom contains eleven electrons distributed in three shells, which hold (reading outward from the atomic center) 2, 8, and 1 electrons respectively. The potassium atom contains nineteen electrons distributed in four shells—2, 8, 8, 1. In either case it is the outermost electrons that encounter the outermost electrons of other atoms during collisions and upon them, therefore, that chemical properties depend. It is the similarity in distribution of electrons that makes the two elements so alike.

But other elements may have similar electron distributions, too, so that sodium and potassium need not be the only alkali metals.

In 1817, for instance, Swedish chemist, Johan August Arfwedson, was analyzing a newly discovered mineral called "petalite." He obtained something from it which, from its properties, he thought might be sodium sulfate. The assumption that it was, however, raised the sum of the elements he had isolated to 105 per cent of the weight of the mineral. He had located an element that was like sodium in its properties but that had to be lighter.

It was a new alkali metal, and because it came from a rocky mineral instead of from plant sources it was named "lithium," from the Greek word for "rock."

The atom of lithium has only three electrons, distributed in two shells—2 and 1.

In the 1850s, chemists developed the technique of identifying elements by heating them till they glowed and then measuring the wavelengths of the light produced. Each element produced a series of wavelengths that could be separated and measured in a spectroscope, and no two elements produced identical wavelengths. Once you identified the wavelengths produced by every known element, you could be sure that any wavelength that was not on the list represented an element hitherto undiscovered.

The first to make use of spectroscopy for the purpose of mineral analysis were the German chemist Robert Wilhelm Bunsen and the German physicist Gustav Robert Kirchhoff. In 1860, Bunsen and Kirchhoff heated material obtained from a certain mineral water known to contain compounds of sodium, potassium, and lithium. They noted a bright blue line with a wavelength not matching those produced by any known element. It meant a new element and they called it "cesium" from the Latin word *caesius* meaning "sky blue." (The British call the element "caesium," which is closer to the Latin but which adds only a misguiding letter as far as pronunciation is concerned.)

A few months later, in 1861, Bunsen and Kirchhoff analyzed a mineral known as "lepidolite" and discovered a deep-red line that was not on the lists. Another new element—and they named it "rubidium" from the Latin word for "deep red."

As it turned out, rubidium and cesium were two new alkali metals. The rubidium atom contains thirty-seven electrons in five shells—2, 8, 18, 8, 1—while the cesium atom contains fifty-five electrons in six shells—2, 8, 18, 18, 8, 1. The five alkali metals are summarized in Table 1.

Table 1

Alkali metal	Total Electrons atomic number	Electron distribution
Lithium	3	2, 1
Sodium	11	2, 8, 1
Potassium	19	2, 8, 8, 1
Rubidium	37	2, 8, 18, 8, 1
Cesium	55	2, 8, 18, 18, 8, 1

That one electron in the outermost shell explains just about everything about the alkali metals. The outermost electron (negatively charged) is held very weakly by the positively charged nucleus at the atomic center, so it is easily removed, leaving a singly charged positive atomic fragment ("ion") behind. The alkali elements are therefore easier to "ionize" than any other metals are.

What's more, as we go up to the scale of atomic number, that outermost electron becomes easier and easier to remove since there are more and more inner electrons to insulate that outermost one from the nucleus. Cesium is therefore easier to ionize than any of the other elements in Table 1.

This is interesting in connection with an "ionic drive" for spaceships. To get out of the atmosphere and away from the immediate neighborhood of a planet requires the brute force of a chemical reaction. Once out in free space, however, instead of using heat to hurl exhaust gases out a nozzle at moderate speeds, you can use an electromagnetic field to flip out ions at nearly the speed of light. Since ions are so light, the acceleration is only gradual, but it mounts up. Cesium, as the easiest substance to ionize, is the most efficient material for an ion drive, and one gram of cesium, converted into ions and emitted, will slowly build up an acceleration at least 140 times as great as that produced by the burning of one gram of any known fuel.

The facts that the outermost alkali-metal electron is so easily removed and that chemical reactions involve electron transfers mean that the alkali metals undergo reactions particularly readily. They are chemically active substances. For instance, they are among the few substances that react with water, tearing away the oxygen atoms in the water molecule, combining with those oxygen molecules and liberating free hydrogen.

The reactivity increases with atomic number. Thus, lithium reacts with water in rather sober fashion, but sodium does so more vigorously. Sodium liberates hydrogen and heat in such quantities that ignition becomes a simple matter and a "sodium fire" is all too easy to bring about.

Despite this, sodium metal is used as a reagent in organic chemistry (for instance, to dry organic liquids, since the sodium will not react with the organic liquid but *will* combine with and remove every last trace of water present). When I took graduate organic chemistry, the class was warned in detail about the possibility of sodium fires, and the promise was made that anyone who started one would be permanently dropped from the class. I paled at the threat because I knew perfectly well that if there was even *one* sodium fire in the class, it would be mine. —Fortunately, there were none at all and I survived.

Potassium reacts so vigorously with water that the heat of the action ignites the hydrogen at once. Rubidium is even more active, and cesium explodes on contact with water and will react even with ice at temperatures down to —116° C. Rubidium and cesium will also react rapidly with oxygen and catch fire spontaneously on exposure to air.

The alkali-metal hydroxides are more alkaline than the carbonates, and the alkalinity increases as one goes down the list. Thus, the most alkaline is cesium hydroxide, which is so reactive that it will chew up even such ordinarily resistant substances as glass and carbon dioxide. It must be stored in silver or platinum containers out of contact with air.

The outermost electron of the alkali-metal atoms has interesting consequences even when it stays where it belongs. It is held so weakly by the central nucleus that, compared with other electrons of its own and other atoms, it bellies out and takes up an extraordinary amount of room. This means that when the alkali-metal atoms come together to form a solid chunk of matter, the atoms are widely spaced and there are relatively few nuclei for a given volume. In other words, the alkali metals possess unusually small densities; see Table 2.

Table 2

Alkali metal	Density (grams per cubic centimeter)
Lithium	0.534
Sodium	0.971
Potassium	0.862
Rubidium	1.532
Cesium	1.873

In general, density depends both upon the mass of the individual atomic nuclei and on the arrangement of those nuclei. The mass of the nucleus goes up with the atomic number so that the density rises as one goes down the list of alkali metals. A more open arrangement of atoms allows potassium to be less dense than sodium.

Even the densest of the metals listed in Table 2 is only a little denser than magnesium, which is the least dense of metals capable of being used in construction (alkali metals are not). Rubidium is well under magnesium's mark of 1.738 grams per cubic centimeter, and the remaining alkali metals are still less dense.

The density of water is 1.000 grams per cubic centimeter, so that lithium, sodium, and potassium are less dense than water and would float in it. (In fact, the poor chemistry student who inadvertently tosses some waste sodium into the sink will, in the few moments before the fire starts and he sees his career shaken to its foundations, watch the little bits of silvery metal hissing, spinning, and floating on the water.)

Sodium and potassium are generally kept under kerosene, for safety's sake. Kerosene, made up of inert hydrocarbon molecules, has a density of about 0.75 grams per cubic centimeter so that sodium and potassium sink in kerosene and rest comfortably at the bottom of the bottle. Lithium, less dense still, would float in kerosene. It is only slightly more than half as dense as water and is the least dense of all metals.

That bulky outermost electron is easily pushed inward (well, comparatively) so that all alkali metals are unusually compressible for solids. Cesium is the most compressible of all those we have considered.

The feebly held outermost electron can easily drift from atom to atom, which is what is required for conducting heat and electricity. The alkali metals do very well in this respect but are exceeded by copper, silver, and gold, which also have a single electron in an outermost shell—but under somewhat different conditions that allow those elements to remain nonreactive and dense.

The outermost electron, all alone and feebly held, performs the function of binding two neighboring atoms only indifferently well. The atoms don't remain in place very rigidly, and the alkali metals are, therefore, soft and waxy, lithium the least so. In using sodium metal in the laboratory, I well remember squeezing it through a "sodium press" by muscle power, and watching it squirt out like rather stiff toothpaste.

Although the single outermost electron holds atoms together well enough to make the alkali metals solid at 25° C, they do so only barely. The alkali metals, as a class, are low-melting; see Table 3.

Table 3

Alkali metal	Melting point (° C)
Lithium	179
Sodium	97.8
Potassium	63.6
Rubidium	38.9
Cesium	28.5

The melting point goes down as the atomic number goes up and the outermost electron is held more and more weakly. By the time you reach rubidium, you have a melting point at a temperature only 9.1 degrees above that which melts gallium. As for cesium, that melts at a temperature 1.3 degrees *below* that which melts gallium. Of all the metals whose melting points have been measured, it ranks lowest after mercury. It melts at a temperature only 3.5 degrees above 25° C, and in terms of Fahrenheit its melting point is only 83.3°.

Yet, by the standard I have set, cesium is solid and it would seem that mercury and bromine remain the only two liquid elements.

But wait, who says that there are only five alkali metals? If one works up the list of elements from cesium, one comes eventually to an element with an atomic weight of 87; one with eighty-seven electrons in its nucleus, distributed as follows: 2, 8, 18, 32, 18, 8, 1. It, too, is an alkali metal—lying beyond cesium.

Element number 87 was not discovered until 1939, when it was detected by the French chemist Marguerite Perey, who was purifying a sample of the radioactive element actinium. The new element was named "francium" after Perey's native land.

As it happens, francium is not a stable element. Not one of its known isotopes is stable and there is no chance that there exists an unknown one that could be stable. The least unstable isotope is francium-223, and that has a half-life of not more than twenty-one minutes. This means that only minute traces of francium can possibly exist on Earth, and only minute traces of francium can possibly be created in the laboratory. The very concept of sizable chunks of solid francium is unrealistic, for its rate of breakdown would yield enough energy to vaporize any reasonable piece.

Consequently, we cannot work out the physical properties of francium metal by direct observation and it is never listed along with the other alkali metals when data is given in the tables in this article.

Yet we can work out its properties by analogy. If we could pretend that francium was a stable alkali metal, we could say, with considerable assurance, that it would be even more active than cesium, that it would explode on contact with water, and so on. We could be sure that francium hydroxide would be more alkaline than cesium hydroxide, that francium would be a good conductor of heat and electricity, that it would ionize more readily than cesium, be softer and more compressible, that it would form a whole list of compounds analogous to those of the other alkali metals, and so on.

And what about the melting point of francium? In Table 3 we saw that the melting point goes down as the alkali-metal atomic number goes up. From lithium to sodium the drop is 81.2 degrees; from sodium to potassium it is 34.2 degrees; from potassium to rubidium it is 24.7 degrees; from rubidium to cesium it is 10.4 degrees. It is fair to suppose that the drop from cesium to francium would be at least five degrees.

In that case, the melting point of francium is quite likely to be 23° C, and by the arbitrary standard we have been using it would be considered a liquid—the third liquid element.

And is that all, or can we still go on beyond francium in the list of elements? So far, all the elements up to 105 either occur in nature or have been prepared in the laboratory, and chemists are trying to form atoms of still higher atomic numbers. We would have to reach element number 119, however, to find a seventh alkali metal.

We might call element 119 "ekafrancium" (*eka* is the Sanskrit word for "one" and it is routinely used for an undiscovered element one beyond a particular analog). Ekafrancium would have an electron arrangement of 2, 8, 18, 32, 32, 18, 8, 1; and it would have all the alkali-metal properties in even more advanced degree than cesium and francium. It would certainly be a fourth liquid element if its undoubtedly enormous instability were ignored.*

---

* Element 118 would be a seventh noble gas, "ekaradon." Its boiling point would be about —20° C, and it would be a twelfth gaseous element. Just thought I'd mention it.

93

7

All Gall

My wife, Janet, who is a physician, has troubles with me. She is extremely diet-conscious while I myself, having always enjoyed an iron digestion, ask only that the portions be healthy in size and worry about the healthy-in-content afterward if at all.*

She is therefore more than a little irritated over the fact that one of my current projects is the writing of a rather large book on diet and nutrition. This came up, particularly, once when we were having brunch at the house of my brother, Stan.

My sister-in-law, Ruth, having dished up a delightful and complicated dish, at the base of which was scrambled eggs, reserved for herself a portion approximately the size of the end joint of the little finger.

Janet looked at her inquiringly. "Is that all you want?" she asked.

"I keep thinking of cholesterol," said Ruth.

At once Janet turned to me and, with an air of loving concern, said to me, plaintively, "Why don't *you* think of cholesterol, Isaac?"

"I do," I said squaring my elbow and preparing to dig into my plateful. "I think of it all the time. I *love* it."

Whereupon Janet said with a sigh, "How can they possibly ask you to write a book on diet and nutrition?"

And my brother, Stan, said, with a grin, "Like asking Hitler to write a history of the Jews."

---

* No, I am *not* fat. I weigh 180 pounds, just enough to lend me an air of attractive resilience.

After that, what can I do but write an article on cholesterol?

The story of cholesterol starts with the liver, an organ characteristic of vertebrates and not found in any other form of life. It is the largest gland in the human body, weighing three to four pounds, and is the most important chemical factory of the body. Among other things, it secretes a juice which flows through a duct into the first part of the small intestine, where it mixes with the food that has come churning out of the stomach.

The liver secretion possesses no enzyme and does not directly digest any portion of the food. It does, however, contain substances with detergent properties that accelerate the breakup of fat globules in the food into tiny droplets. This makes it easier for fat-digesting enzymes secreted by other glands to attack the fat.

The secretion formed by the liver is called "gall" from an old Anglo-Saxon word for "yellow" since the fresh juice has a yellowish cast. The secretion is also called "bile" from a Latin word of uncertain derivation. The Greek term for the secretion is *chole*. All three terms, Anglo-Saxon, Latin, and Greek, find their way into English words having some connection with the secretion. For instance, the duct through which the secretion passes is called the "bile duct."

The ancient Greek physicians considered two varieties of bile to exist, "black bile" and "yellow bile." In this, they were wrong, for there is only one bile, though it may be different in color, depending on the state of its freshness.

The ancient medical theories had it that when anyone suffered an overproduction of black bile, he was given to sadness and was "melancholic" (from a Greek phrase meaning "black bile"). An overproduction of yellow bile meant a person was given to anger and was "choleric." Note *chole* in both words. Both "bile" and "gall" are also used to refer to human emotions such as rage and rancor.

The liver manufactures about half a liter (one pint) of bile a day. It doesn't pour it into the small intestine continuously, however; that would be wasteful. The bile is vented only when the food enters the small intestine.

In between meals, the bile that is formed is stored in a special sac called the "gall bladder," a pear-shaped organ about five to seven centimeters (two to three inches) in length. The gall bladder has a capacity of about fifty milliliters (three cubic inches.)

Once bile is stored, water is reabsorbed through the walls of the gall bladder so that the bile grows steadily more concentrated as it waits to be used. At maximum, it is ten to twelve times as concentrated as the original bile and can therefore hold the active ingredients of over a day's supply of the material. As food enters the small intestine, the walls of the gall bladder contract and the concentrated bile squeezes into the small intestine.

Among the ingredients of the bile, there are three substances that have the potential of causing trouble: (1) calcium salts, (2) bilirubin, a pigment which gives bile its color, and (3) cholesterol. All three are relatively insoluble and all three remain in solution in the bile only with difficulty. As the water is withdrawn from the bile stored in the gall bladder, the difficulty of keeping those compounds in solution increases.

In some cases, one or two or all three precipitate out of solution to form "gallstones."† The chances of forming gallstones seem to be connected with sexual factors (they are more common in women than in men), with hereditary factors (they are more common in fair people than in dark people, more common in Jews than in Japanese), and with diet (more common in fat people than in thin people). Also, since fat in food apparently stimulates the flow of bile and increases the chance of stone formation, gallstones are more common in

† This chance fact gives me the opportunity of observing, with my usual brilliance, that "All gall is divided into three parts" and hands me the title for the essay. If, however, you've never heard of Julius Caesar's remark about Gaul, all this brilliance is wasted on you.

those with a high-fat diet than in those with a low-fat diet.

The gallstones may be very tiny, almost like a fine grit, or may be so large that one will almost fill the gall bladder. Usually, they are pea-sized. They become particularly troublesome when they block the duct leading from the gall bladder, in which case they can produce severe abdominal pain and, eventually, can damage the liver. Often, the best treatment is to remove the gall bladder. This hampers the efficiency of bile function but not enough to prevent people from living perfectly normal lives minus their gall bladder.

From the chemical standpoint, gallstones are associated with the French chemist Michel Eugène Chevreul, who is remarkable for being the longest-lived of the first-magnitude scientists. Born in 1786, he died in 1889 at the age of 102.6 years. He was still an active scientist in his tenth decade and, indeed, founded gerontology, the study of old age, using (who else?) himself as his subject. His hundredth birthday was celebrated by chemists everywhere with terrific enthusiasm and he was hailed as "the Nestor of science."

What concerns us, however, is the fact that in 1823, while he was still a mere toddler in his thirties, he investigated gallstones and isolated from them a fatty substance of a pearly appearance. He thought it was solidified bile, so he called it "cholesterin" from Greek words meaning "solid bile."

It took over a century for the chemical structure of the cholesterin molecule to be completely worked out. The molecule is made up of seventy-four atoms, of which twenty-seven are carbon atoms arranged in four connected rings and three side chains. Two of the side chains possess one carbon atom each and the third contains eight carbon atoms.

Attached to the twenty-seven carbon atoms are a total of forty-five hydrogen atoms and one oxygen-hydrogen combination (a hydroxyl group). The hydroxyl group is characteristic of alcohols, and to the name of alcohols the suffix "-ol" is commonly added. Hence, as the structure of cholesterin became better known, its name was altered to "cholesterol" and that is how it is known today.

Cholesterol is one of a family of compounds, all with the same four-ring system and the attached hydroxyl group. Together, they are known as "sterols." As far as we know, every plant and animal cell, whether a unicellular organism or part of a multicellular one, contains sterols. Clearly, they are essential to the cellular machinery; we can see that. The only trifling catch to this is that we don't know exactly *why* they are essential to the cellular machinery. No one has yet established exactly what they do.

Although plant cells and animal cells both have sterols, they don't have the same sterols. All animal cells contain cholesterol, but no plant cell does, and there are no known exceptions to this rule. An example of a plant sterol is stigmasterol, which differs from cholesterol in having a ten-carbon side chain in place of the eight-carbon one. Ergosterol, which occurs in one-celled plants such as yeast, splits the difference, and has a nine-carbon side chain. Whereas cholesterol contains twenty-seven carbon atoms, ergosterol has twenty-eight and stigmasterol twenty-nine.

As far as we know, no animal cell lacks the ability to make its own cholesterol out of very simple two-carbon fragments, universally present in any cell that is not actually starving to death. This means that no animal must depend on its diet as a source of cholesterol.

In particular, human beings do not need cholesterol in the diet. Human cells can make their own. To see the significance of this fact, let's start over.

Plants can make all the components of their cellular tissue out of the simple molecules of soil and air. They have to, since that's all they get to start with, and any particular plant that couldn't be entirely self-sufficient would have to die.

Animals, however, which eat plants (or which eat animals which eat plants, or which eat animals which eat animals which eat plants—however many animals in the food chain, it always comes down to plants in the end) obtain a variety of complicated molecules in their food. For the most part, they break these down to simpler fragments and rebuild their own varieties of complicated molecules.

A number of animals have, however, lost the ability to manufacture all the different atom combinations necessary to their functioning out of the simple breakdown products of food materials. Some atom combinations must be extracted from the food intact, must be absorbed and utilized in that form. If the atom combinations are not present in the food, the animal must do without; if it cannot do without, it must die.

These dietarily essential compounds must fulfill two criteria. For one thing they must contain unusual atom combinations not present in sizable quantity in other tissue components. If this were not so, the dietarily essential material could be formed from the other components and it would not be dietarily essential.

Secondly, the dietarily essential compounds must be needed in relatively small quantities, as otherwise the organism would be taking too great a risk to rely on its being present in food quantities sufficient for need.

This sounds as though I think organisms are acting purposefully in organizing their chemical properties, but that's not it, of course. The blind path of evolution is sufficient. If, through chance, a particular organism is born with a dietary need for something it requires in large quantities, the chances are it won't get enough and it will die. Only those organisms born, by chance, with reasonable dietary requirements, make it.

But why have any special dietary needs at all? Wouldn't a cell or an organism be better off if it made all its own atom combinations out of the simplest possible materials, and depended on its diet only for those substances invariably present everywhere?—Not necessarily.

To be completely self-sufficient would mean the loading down of each cell with quantities of chemical machinery designed to manufacture all potentially useful atom groupings. By clearing out the machinery for those groupings needed in minor amounts and relying on the diet for those, room is made, so to speak, for other machinery more necessary to the complex physiological functioning of advanced animal organs such as the brain. In short, we trade a little clumsiness in the diet for virtuosity in other directions.

These dietarily essential substances are very noticeable to us, since it is possible to lack them and be in trouble, whereas we can't lack the substances we can manufacture at will (short of starvation). We therefore get the idea that the dietarily essential substances are *particularly* important. We even drop the reference to the diet and speak of "essential fatty acids," "essential amino acids," "essential vitamins," and so on. We get the idea that everything else is nonessential.

Quite the reverse. These vitamins and other substances are indeed essential. Within the body, however, are even more essential substances, substances so essential that they dare not be left to the diet, and so do not impinge upon our attention.

In other words, the mere fact you don't need cholesterol in the diet shows how important cholesterol is to your body chemistry.

And there's convenience there. If you're a complete vegetarian and eat no animal products (not only no meat, but no fish, no butter, no milk, no eggs) then you have no cholesterol in your diet; none at all. Nevertheless your body does not suffer. It can break down any sterols it does get into two-carbon fragments, throw them into the general hopper of such fragments and then, out of them, build all the cholesterol it needs. It could do this even if it got no sterols of any kind to

100

begin with since it can use two-carbon fragments obtained from other dietary components.

If a person *does* eat animal food, the cholesterol it contains is absorbed (rather inefficiently) and is added to the body's supply directly.

Since cholesterol is water-insoluble and fat-soluble (these two properties almost invariably go together in the case of carbon-containing compounds), it is found in the fatty portions of animal foods—in egg yolk, in milk fat (and therefore in cream, butter, cream cheese, and so on), in the marbling of the meat.

If, for any reason, you, like my sister-in-law, Ruth, wish to go easy on cholesterol, you eschew eggs, cream, butter, bacon; you trim the fat off your meat; and so on. If you really want to cut it all the way, you become a strict vegetarian.

Why be concerned to lower the cholesterol in the diet? Obviously because there is some desire not to encourage the body to make too much of it. Ideally, the body should maintain a balance, and make less of its own as more floods in from the intestines. Unfortunately, the ideal isn't always attained. If the body's cholesterol-synthesizing machinery doesn't respond perfectly, an unusual amount of cholesterol in the diet may result in an unusual amount of cholesterol in the body.

The total cholesterol content of a man weighing seventy-seven kilograms (170 pounds) is 250 grams (8¾ ounces.) The body, in other words is about one third of one per cent cholesterol. It's not really a very minor constituent.

Tissues generally can form their own cholesterol. The chief cholesterol-former for the body as a whole, however, is the liver. The liver secretes cholesterol into the blood stream so that it is available wherever needed.

The quantity of cholesterol in the blood is about 200 miligrams for every hundred milliliters, the normal range fluctuating from 150 to 250. The blood is thus about one fifth of one per cent cholesterol, rather poorer in it than is the body generally.

The cholesterol of the body is the raw material for a number of important compounds chemically related to it and therefore called "steroids" ("sterol-like"). The hormones of the adrenal cortex—cortisone, for instance—are steroids. The sex hormones are steroids. In the bile there are "bile acids" that are steroids, and so on. All these steroids make up a minor portion of the body, however. Most of cholesterol remains cholesterol and is used as such.

The portion of the body that is richest in cholesterol is the nervous system. There we can encounter masses of nerve cells which, in bulk, have a grayish appearance and therefore referred to as "gray matter."

The nerve cells have fibers extending outward in all directions, and generally have one particularly long fiber called the "axon." The axon is sheathed in fatty insulation, and that portion of the brain which consists of massed axons is called, from the white appearance of the fat, "white matter."

Nerve cells work by producing tiny electric potentials which travel along the various fibers, particularly the axons, as "nerve impulses." These impulses manage to jump from fiber to fiber across narrow gaps, and the entire working of the nervous system depends on those electric phenomena not leaking away.

Animals such as squids solve the problem by making the axon quite thick so as to lower the resistance and make it very easy for the tiny impulses to stay in the axon rather than to stray out of it.

Among the vertebrates, the problem is solved by having the axons surrounded by the fatty insulation I've already mentioned, a so-called "myelin sheath." The myelin sheath probably acts in an insulating fashion, but that can't be its only function. If it were just a matter of insulation, simple fat molecules would do. Instead, the myelin sheath contains quite complex fatlike molecules, including a number that do not appear in noticeable amounts outside the nervous system.

The myelin sheath may serve to maintain the integrity of the axon, which sometimes stretches so far from the cell proper that one could scarcely expect that cell to control and co-ordinate all those far-distant molecules. Then, too, the myelin sheath must somehow act to increase the speed of the nerve impulse. In general, the thinner the axon, the greater the resistance and the slower the impulse. With a sheath, however, the nerve impulse moves along the thin vertebrate axon with remarkable speed. It can pass from end to end of a tall man in a fiftieth of a second.

Among the compounds present in the myelin sheath is cholesterol. Cholesterol makes up about one per cent of the weight of gray matter, and four per cent of the weight of white matter, since two out of every five molecules in the myelin sheath are cholesterol. Since we don't know exactly what the myelin sheath does, we certainly don't know what the cholesterol does to contribute to its working.

Still, there is no doubt that whatever cholesterol does, it is essential to the myelin sheath, that the myelin sheath is essential to the nervous system, and that the nervous system is essential to us.

So far, then, cholesterol seems like a good guy and there is every tendency to clap the liver on its back and say, "Go ahead, make a lot."

There is a tendency, however, for cholesterol and other fatty materials to settle out of the blood and onto the inner walls of the medium and large arteries.

When this happens, the walls are thickened and hardened. In general, such a condition is called "hardening of the arteries," or "arteriosclerosis," which is the same phrase in Greek. The particular kind of hardening which results from the deposition of fatty substances on the inner walls is called "atherosclerosis." The prefix is from a Greek word for "meal," because the fatty deposits have a mealy appearance.

The atherosclerotic deposits can lead to a variety of very bad results. Since the artery wall is thickened, the bore is narrowed so that less blood can be carried. Vital organs can be starved for oxygen and this has its worst effect when the heart is starved. The heart can't afford to be starved, and when its coronary arteries become atherosclerotic, the agonizing pains of angina pectoris result.

Then, too, the thickened arterial walls lose flexibility. Ordinarily, when the heart contracts and a stream of blood slams in the main arteries, the walls expand, making room for the blood and reducing the pressure against themselves. With the loss of flexibility, the arteries do not expand as much as they should and the blood pressure goes up. Under the battering of the high blood pressure, the walls of the arteries undergo further degenerative changes and some may even rupture.

Again, the atherosclerotic layer has a rough surface as opposed to the smooth surface of the healthy arterial wall. The roughness encourages clot-formation and once in a while such a clot may break loose and go swirling along the blood stream. It may then, just possibly, lodge in one of the smaller arteries in such a way as to block the blood stream through that artery altogether.

When this happens in one of the arteries leading to the heart muscle, you have "coronary thrombosis," the familiar "heart attack." The portion of the heart muscle fed by that artery dies and life can continue (subject to the probability of other attacks for the same reason), but if the blockage is serious enough, death can follow quickly.

If the clot blocks one of the arteries in the brain, you experience a stroke. Here again, death can follow quickly, or a portion of the brain will die and the patient will survive. However, partial brain death is more serious than partial heart death, for the former will lead to paralysis of one sort or another, permanent or, if other parts of the brain can take over, temporary.

Atherosclerosis and allied circulatory disorders are becoming an ever-more-important cause of death. They cause more deaths than all other causes combined.

Nor is it entirely a question of medical success in

other directions. True, many diseases have been brought under control so that people nowadays avoid dying of diphtheria and typhoid fever and pneumonia and appendicitis, and can live on till their circulatory system fails. That, however, isn't the whole answer. More and more comparatively young people are dying of circulatory disorders.

One hint lies in the fact that these disorders are a disease of the prosperous. Atherosclerosis is more common in affluent nations than in poor ones and, within any nation, more common among the affluent classes.

One fact of prosperity is that there is a diet that goes along with it. Prosperous people eat more food than poor people do, and, even more important, they eat more expensive food—which means more animal food.

Eating animal food means eating cholesterol and, since cholesterol is found in the atherosclerotic deposits, there might be a connection. As long ago as 1914, it was found that when rabbits were fed large amounts of cholesterol, atherosclerosis could be induced. (Rabbits in the ordinary course of nature never eat cholesterol since they are herbivores, and their body mechanism is ill-equipped to withstand this abnormal dietary flood.)

More and more studies have shown that with high dietary cholesterol, there is a tendency to produce a level of cholesterol in the blood that is higher than normal, and, therefore, an increase in the tendency to atherosclerosis.

If this were all there was to it, the matter would be simple. There are, as it happens, other factors. Some saturated fatty acids, when present in the diet, tend to increase blood cholesterol levels, while some unsaturated fatty acids tend to decrease it.

Then, too, there are factors that encourage atherosclerosis even where cholesterol is not directly involved. Smoking is an example. There are three times as many heart attacks among smokers as among nonsmokers, and the incidence of fatal heart attacks is five times as high. (Consequently, any smoker who cuts down on dietary cholesterol while continuing to smoke—and feels virtuous about it—is a jackass.)

Hereditary factors have their effect, too, as do life styles, psychological attitudes, and so on.

There are no simple ways of absolutely avoiding atherosclerosis and thus increasing the chances you will live long. But you can try several strategies. You can eliminate the tobacco habit if you're an addict. You can indulge in moderate exercise, avoid stressful situations, cultivate a serene disposition, and, if you are overweight, reduce.

Also, you can cut down the cholesterol intake if you're overdoing it. It isn't so difficult to trim the fat off the meat, and to use leaner cuts in the first place, or to reduce the intake of eggs, butter, and bacon, and—

(Good heavens, Janet, I may be talking myself into it.)

8

The Smell of Electricity

Not long ago, I had occasion to drive through the town of Fulton, New York, on my way to Oswego, where I was scheduled to give a talk. And there I stopped my car and cried out, "Chocolate!"—For chocolate is a mad passion of mine.

It seems there is a chocolate factory in the town, and it took a while for my wife to persuade me to put the car in gear again and start driving, before my eager snuffling made me too oxygen-drunk to be trusted at the wheel.

I was still under the spell of the memory at dinner that night, which I shared with a few students and professors. I told them of my experience and, recalling another mad passion of mine, said, "Yes, indeed, I am quite sure that if there were a heaven and if I were judged worthy of entrance, my reward for a well-spent life would be chocolate-covered girls."

At least one of the professors present was struck by the thought for he kept muttering, "Chocolate-covered girls! Hmm!" for the rest of the evening. Perhaps he was trying to assess what a redhead dipped in semisweet would be like.

But if no other odor is quite that of chocolate (at least to me) all odors are evocative. In my incarnation as professional chemist, I was accustomed to odors of quite another variety and, while most were not pleasant in themselves, they bring back delightful memories of every one of the now-very-rare times (alas!) that I walk into a chem lab.

So let's talk about smells.

In the middle and late 1700s, scientists were fascinated by electricity. They stored electric charges in Leyden jars, then discharged them, watched the sparks, listened to the crackles, felt the shocks, and had as much fun as you and I would have at a carnival. And sometimes, they detected a funny smell that seemed characteristic of such apparatus.

The smell was specifically referred to for the first time (as far as we know) in 1785 by a Dutch physicist, Martin van Marum, who built giant friction machines out of which to draw nice, fat sparks.

It wasn't till 1839, however, that someone, catching the smell, thought of it not as just the smell of electricity, but as a definite chemical compound. That someone was a German chemist, Christian Friedrich Schönbein, who had the two necessary ingredients in his laboratory, electrical equipment and poor ventilation.

By 1839, the element chlorine had been discovered, and the smell of electricity was somewhat like chlorine so that Schönbein couldn't help but think he had a chlorinelike substance, perhaps chlorine in combination with other elements. Whatever it was, he named the substance "ozone" from the Greek *ozo* meaning "I smell."

One thing one might do would be to try to find out under what kind of chemical conditions the smell appeared. In other words, suppose the electrical equipment was sparkling clean and was surrounded by pure nitrogen, or pure hydrogen, and electric sparks were discharged through the gas. Would the smell appear? (Answer: No.)

In 1845, two Swiss scientists, a chemist named Jean Charles Gallisard de Marignac and a physicist named Auguste Arthur de la Rive, passed pure dry oxygen through an electric discharge and, behold, the smell appeared. Whatever ozone might be, it had to be a form of oxygen since there was nothing else out of which it could be formed.

Trying to work out what the form of oxygen might be was a problem. One trouble was that chemists, through the first half of the 1800s, were not sure about the manner in which atoms grouped together to form molecules. Nowadays we know that ordinary oxygen is made up of molecules that each contain two oxygen atoms, so that "molecular oxygen," as it occurs in the atmosphere, for instance, is written O_2. The evidence for that, however, wasn't straightened out and made plain till 1858, when the Italian chemist Stanislao Cannizzaro finally showed how to determine molecular weights from vapor densities.

There was no way at that time of collecting enough pure ozone to measure its density, but there were other lines of attack. Gases diffuse. Their molecules blunder their way past other molecules and even through tiny holes in apparently solid materials, so that if you uncork a bottle of strong-smelling material at one end of a room, you will sooner or later smell it at the other end even if there is, let us say, a cardboard partition between.

The speed of diffusion is inversely proportional to molecular weight. That is, a more massive molecule moves more slowly than a less massive molecule. That makes sense, of course, but the trick is to determine just exactly how much more slowly the molecules move as the mass increases.

In 1834, the Scottish chemist Thomas Graham, working with molecules whose comparative masses he felt reasonably certain he knew, made careful measurements and decided that the change varies as the square root of the mass. In other words, if a particular molecule moves at a certain rate, one that is four times as heavy will move two times as slowly (or one-half as fast, if you prefer); one that is nine times as heavy will move three times as slowly and so on.

This relationship (called "Graham's Law") was found to work very well and could be used with confidence once the structure of molecules was straightened out. For instance, one could study the rate at which ozone diffused. For that, one didn't have to collect quantities of pure material; one could use merely traces and note when some detectable chemical property could be found at some given distance from the starting point.

In 1868, the chemist J. Louis Soret ran experiments designed to compare the rate of diffusion of ozone with that of carbon dioxide and chlorine. It turned out that ozone diffuses about five per cent more slowly than carbon dioxide and about twenty-two per cent faster than chlorine. The molecular weights of carbon dioxide and chlorine (44 and 71, respectively) were known and it could therefore be calculated that ozone has a molecular weight of 48. Since the oxygen atom has a weight of 16, it was clear that ozone must be made up of three-atom molecules. Whereas ordinary oxygen is O_2, ozone is O_3.

In 1922, the German chemist Georg Maria Schwab produced pure ozone for the first time and was able to measure its density—which confirmed the O_3 formula.

Ozone is not merely a form of oxygen. The two are distinctly different substances that just happen each to be made up of oxygen atoms exclusively.

We are not surprised that carbon dioxide (CO_2) is a compound that differs radically from carbon monoxide (CO), even though the only difference is an extra oxygen atom in the molecule of the former. Substitute an oxygen atom for the carbon atom in each and we have ozone (OO_2) and oxygen (OO).

The difference can be seen in many ways. Oxygen is a colorless gas that can be made to condense, at very low temperatures, into a pale blue liquid, and then to freeze, at lower temperatures, into a darker blue solid. Ozone is a pale blue gas that can be made to condense into a deep blue liquid and to freeze into a solid so deeply violet in color as to be virtually black.

Both oxygen gas and ozone gas contain the same number of molecules in a given volume. The fact, however, that the individual ozone molecule has three oxygen atoms to two for the oxygen molecule, makes ozone gas half again as dense as oxygen gas. A liter of oxygen gas under ordinary atmospheric conditions weighs 1.43 grams; a liter of ozone gas under the same conditions weighs 2.14 grams.

The greater density persists in the liquid form. At its boiling point, the density of liquid oxygen is 1.142 grams per milliliter (or 1142 grams per liter) which is about 800 times as dense as the gas. At the same temperature, liquid ozone has a density of 1.571 grams per milliliter, which is about 750 times as dense as the gas. (The three-atom molecules can't pack together quite as tightly in liquid form as the two-atom molecules can.)

All things being equal, substances with large massive molecules tend to have higher boiling points and freezing points than those with small molecules.

Liquid oxygen freezes to the solid form at— 218.8°C, or 54.4 degrees above the absolute zero—which we can write as 54.4° K—and boils at 90.2° K. Liquid ozone, however, with its larger molecule, freezes at 80.5° K and boils at 161.3° K.

Ozone is also considerably more soluble in water than oxygen is. At 0° C a liter of water will dissolve 4.9 cubic centimeters of oxygen but will dissolve 49 centimeters of ozone, just ten times as much of the latter.

You might think that liquid oxygen and liquid ozone, both consisting of oxygen atoms only, would be at least enough alike to mix freely, but that is not so. In the liquid oxygen temperature range, one part of liquid oxygen will mix with three parts of liquid ozone, and vice versa. However, if you were to mix an equal proportion of liquid oxygen and liquid ozone and were to stir well, you would end with two separate liquids with a clear dividing line. The top liquid, a deep blue in color, would be chiefly liquid oxygen, with some liquid ozone dissolved in it. The bottom liquid, nearly black, would be chiefly liquid ozone, with some liquid oxygen dissolved in it.

Oxygen has no odor. It can't have. We breathe it

constantly; we are thoroughly saturated with it. Whatever chemical changes in the inner linings of our noses produce the sensation of odor, none can take place with oxygen since any possible reaction has already taken place at the very beginning of the smell-sensation in the individual. If we could imagine ourselves as somehow living without oxygen at all and with all gaseous oxygen removed from our body and if we were *then* to breathe a bit of oxygen, we should undoubtedly receive the sensation of a pronounced and, probably, unpleasant smell.

Well, ozone has such a smell, and a strong one, too. In fact, ozone can be just barely detected by smell where there is as little as 0.01 part per million (ppm) in air, provided no other smells are competing.

Ozone is highly poisonous, too (as opposed to oxygen, which is immediately and continually essential to life). A concentration of 0.1 ppm in air is the maximum allowable for eight hours continual exposure. Ozone is about a hundred times as poisonous as carbon monoxide.

Ozone formation from oxygen requires an input of energy. The pair of oxygen atoms forming molecular oxygen occupy a stable position with respect to each other. Left to themselves under ordinary conditions, they jiggle about and bounce off each other, but in doing so, they neither join together as a double molecule nor split apart into single atoms.

To add a third oxygen atom to so comfortably married a couple is not easy. One way of accomplishing it is by adding energy to the system in the form of an electric discharge—the method by which ozone was first discovered.

Another way is by exposing oxygen to light. Not ordinary light, which isn't energetic enough, but ultraviolet light. If oxygen in a quartz container (quartz allows ultraviolet light to pass through itself, where ordinary glass does not) is exposed to ultraviolet, it will smell of ozone when liberated. Then, too, if liquid oxygen is exposed to ultraviolet light (an experiment first carried out in 1907), it grows steadily bluer as liquid ozone is formed.

What probably happens in such cases is the energy of the electric discharge, or of the ultraviolet radiation, will split an occasional oxygen molecule in two, forming free oxygen atoms ("atomic oxygen"). If only atomic oxygen were then present, the atoms would collide and recombine into oxygen molecules, liberating the energy that had gone into the splitting of that molecule (though the energy liberated might well be different in form from that which entered the system to begin with).

However, relatively few of the molecules are split, so that the free oxygen atoms in their blundering are overwhelmingly likely to strike intact oxygen molecules. The chemical activity of a free oxygen atom is extremely high to begin with, and with the added energy of the splitting agent, the atom can attach itself to the oxygen molecule to form ozone.

If an atom adds on to a molecule, thanks to an input of energy, we may expect that it will eventually drop off again with the reliberation of that energy (perhaps in a different form). The more difficulty there is in adding that atom, the greater the ease with which it will drop off.

Ozone, which is formed from oxygen with considerable difficulty, will convert back into oxygen spontaneously with the encouragement of a little heat. The heat causes the ozone molecule to vibrate more energetically and the third atom shakes off. That liberates more energy, which shakes the remaining molecules even more vigorously, producing more breakdowns without the liberated energy ever appearing in a form concentrated enough to allow the re-formation of ozone. Once begun, the process of decomposing the ozone molecules proceeds to completion rapidly. In fact, if care is not taken, the process of decomposition becomes so rapid that the ozone explodes.

Schönbein, who discovered ozone, found that if oxygen containing ozone was passed through a heated tube, it emerged as pure oxygen. That was one of the early experiments that confirmed that ozone was made up of oxygen atoms only.

When ozone decomposes to liberate oxygen atoms, and when nothing else is present, those atoms, lacking the drive of concentrated energy, cannot reattack the oxygen molecules, and can do nothing more than find each other and form oxygen molecules.

Even at room temperature, ozone molecules occasionally break down, but in such small proportions that the occasional union of free oxygen atoms does not deliver much heat. Such heat as is delivered appears at such a low rate that there is time for it to be radiated away into the environment as quickly as delivered and the temperature does not rise. Therefore, although room temperature ozone may very slowly break down, it never does this explosively if pure. Liquid ozone, if pure, will break down so slowly at its low temperature that it might be considered as practically stable.

It may be, however, that there is present some impurity that is more readily attacked by free oxygen atoms than the oxygen molecule itself is. The presence of such substances in ozone increases its instability.

Imagine ozone containing small quantities of molecules made up, at least in part, of carbon and hydrogen atoms. (This is characteristic of any organic molecule of the type that is, or was once, part of living tissue, or that resembles substances that are, or were once, part of living tissue.)

The occasional free oxygen atom produced by spontaneous ozone breakdown even at low temperatures combines readily with carbon or hydrogen atoms and delivers considerable heat. The temperature rises more rapidly, therefore, in the presence of organic molecules than in their absence; breakdown accelerates and quickly reaches the pitch of explosion. Naturally, the greater the concentration of ozone, the more likely this is to happen, so that ozone in high concentration must be treated very carefully, kept very free of impurities other than oxygen, and maintained at a reasonably' low temperature. Otherwise, it becomes an explosion hazard.

It might seem rather surprising that free oxygen, as it occurs in the atmosphere, does as little damage as it does. Oxygen atoms combine rapidly with most other atoms, including, particularly, the carbon and hydrogen atoms in organic molecules. Why doesn't the oxygen in the air combine instantly with all the organic matter in the world (including ourselves) and do it energetically enough to produce one great conflagration that would end with all oxygen gone and all life in ashes?

That this does not happen is entirely due to the fact that the two oxygen atoms in the oxygen molecule hold each other so tightly. As long as they are together, they are relatively harmless, and their combination with other atoms proceeds so slowly as to be virtually nonexistent.

When the temperature goes up, the oxygen molecule vibrates more and more strongly, and the bond between its atoms weakens. There comes a point where the individual oxygen atom will more readily combine with a carbon atom or a hydrogen atom in something organic than remain attached to its twin in the molecule. The combination of oxygen atoms with other atoms releases heat, which further raises the temperature, further weakens the oxygen-oxygen bond and further accelerates combination of oxygen atoms with other atoms.

There is, in other words, an "ignition temperature," and once that is reached, combination with oxygen ("oxidation") continues rapidly, producing in the case of most organic materials vapors hot enough to glow. We have combustion and the appearance of fire.

The third oxgyen atom of ozone is so loosely bound that it takes little or no heat to encourage it to combine elsewhere. Substances are much more likely to combine with oxygen atoms in the presence of ozone than in the presence of oxygen molecules. Ozone is, therefore, a stronger "oxidizing agent" than oxygen is.

The metal mercury, for instance, does not combine

with oxygen at room temperature. It remains shining and metallic in appearance in contact with air. In the presence of ozone, however, mercury rusts and forms an oxide. Silver will rust in the presence of ozone, too, if heated a bit. There are numerous chemical reactions that won't go with oxygen, but will with ozone.

The oxidizing effect of ozone can be put to use in organic chemistry. Here's how it works—

Organic molecules consist of chains or rings of carbon atoms, to each of which other atoms may be attached. Usually, each carbon atom is held to adjacent carbon atoms by virtue of sharing with each neighbor a single pair of electrons. For historical reasons, this is called a "single bond." On occasion the attachment is by way of sharing two pairs of electrons—a "double bond."

In studying the structure of organic molecules, chemists find it important to know if double bonds are present and, if so, where in the structure they are to be found. One way of determining this is to take advantage of the fact that the double bond represents a weak point in the carbon chain.

(You might think that two atoms held by a double bond are more tightly joined than when held by a single bond, but it doesn't work that way. The picture produced by the word "bond" is misleading in this respect. Four electrons crowded between two atoms form a less stable interacting arrangement, and that makes the joining weaker.)

Oxygen itself is not a strong-enough oxidizing agent to take advantage of the double-bond weak point, but ozone is. The ozone molecule can add on rapidlly at the point of the double bond. All three oxygen atoms add on to form an "ozonide." (This process was first reported in 1855 by Schönbein.)

In forming ozonides, chemists use a stream of oxygen in which the concentration of ozone is no more than six to eight per cent in order to avoid too uncomfortably rapid a reaction. The ozonide that is formed is usually itself explosive, so chemists don't let it hang around.

116

They react it with water or other substances, and such a reaction divides the molecule at the point of ozone addition, a division referred to as "ozonolysis."

In place of the original molecule with its double bond, you have, in the case of a carbon chain, two smaller molecules. In the case of a carbon ring that ring is broken and a carbon chain is formed. In either case, by studying the nature of the molecules after ozonolysis, chemists can determine the nature of the original molecule and the exact position of the double bond. Ozonolysis was used, for instance, to determine the structure of the rubber molecule and to direct the art of the chemist toward the formation of artificial rubbers on something better than a hit-and-miss basis.

Sometimes the smaller chemical compound that appears after the breakage of the chain by ozonolysis is more valuable than the original. It is easy, for instance, to get a compound called eugenol from plants. This is easily turned into a related compound called isoeugenol, and this can be broken down by ozonolysis into vanillin, the much more valuable compound that gives vanilla its flavor. This was the most important commercial ozonolysis reaction in the early decades of the twentieth century.

Since then, another ozonolysis has become more important. Oleic acid, the molecules of which contain an eighteen-carbon chain, is universally found in all natural fats and oils. The molecule has a double bond right in the middle of the chain and by ozonolysis it is split into two molecules of nine carbon atoms each, which can then be used as starting materials for certain other substances with useful applications.

Ozone is similar in is chemical reactions to chlorine, since both are oxidizing agents. (In the early days of chemistry, what we call oxidation was so characteristic of oxygen that it didn't seem reasonable to think of it in connection with other substances. Oxidation, however, is brought about by the removal of electrons from the substance being oxidized, and chlorine can accomplish that task. The element fluorine can remove electrons more readily than chlorine, oxygen, or ozone, and is the strongest oxidizing agent known. In fact, fluorine can oxidize oxygen itself by taking electrons away from the oxygen atom.)

Usually colored substances, when oxidized, lose their colors. An agent which will oxidize such substances without seriously affecting the textile material carrying them will serve as a useful bleach. Chlorine and various chlorine-containing compounds serve as bleaches; and so does ozone.

Chlorine will also act to kill micro-organisms. (It will kill us, too, if we breathe enough of it.) This killing action of chlorine is useful in the sterilization of water in swimming pools and in making a city's water supply safe (if not exactly pleasurable) to drink.

Ozonization, less common than chlorination, accomplishes the task more rapidly, and since ozone turns to oxygen in the process, it imparts no bad taste to the water.

Then, too, ozone, if added to the air of cold-storage rooms to an extent of one to three parts per million, will serve a useful purpose. The growth of bacteria and molds, already inhibited by the cold there, will be further inhibited by ozone.

The use of ozone in various purification procedures may have given rise to the thought that ozone is a particularly pure and invigorating form of oxygen in the minds of those innocent of chemistry. Ozone is sometimes used as a synonym for the clean air of the outdoors far away from the city's filth.

As a matter of fact, there is some ozone in the atmosphere that surrounds us, formed there by the action of sunlight. In rural areas, it may reach a level of 0.02 to 0.03 ppm, just enough to smell if the other smells of the countryside didn't drown it out. In cities, there is usually less ozone than that because there is less sunlight—unless there are certain chemical impurities in the air of the type that form smog. That tends to encourage ozone formation by sunlight, and concentrations as high as 0.5 ppm have been reported on smoggy days for short periods of time—a concentration definitely in the danger zone.

Putting aside the effect on human health, the presence of ozone can be troublesome because ozone adds on to double bonds in chemical chains and, in particular, to those in rubber. Ozonized rubber loses its elasticity and becomes brittle, so that smog is hard on automobile tires, which must be specially treated to make them resistant to the effect.

The natural occurrence of ozone in the atmosphere becomes much more important at great heights, and I'll approach that subject from another direction in the next chapter.

9

Silent Victory

I was present not long ago at a very elaborate banquet at which the famous lawyer Louis Nizer delivered one of the two major addresses. It took the form of a skillful, optimistic view of the future of mankind, delivered with perfect eloquence and without notes. It was, in fact, a superlative science fiction oration, and since I was there at the head table with him, I couldn't help but squirm. I was being beaten in my own field—and by an outsider.

Within fifteen minutes after he was done, it was my turn, but I was one of fifty (literally fifty) and I was expected to speak only a couple of minutes. I think it was also expected that I would spend my couple of minutes expressing a humble acknowledgment of my gratitude at the honor being paid me (along with the other forty-nine). However, my talent for humility is poorly developed and there was something else I wanted to do.

I said (speaking rapidly so as to get it all out within the time limit): "Mr. Nizer has given you an excellent picture of a wonderful future, and since I am a science fiction writer, I can't help but envy the clarity and eloquence of his vision. However, we must remember that the various governments of Earth are, in these complex times of ours, the direct mediators of change and it is they who largely determine the nature, quantity, direction, and efficiency of change. We must also remember that most governments are in the hands of lawyers; certainly our own is. The question, then, is what may we expect of lawyers?

120

"And in that connection, there is the story of the physician, the architect, and the lawyer who once, over friendly drinks, were discussing the comparative ancientness of their respective professions.

"The physician said, 'On Adam's first day of existence, the Lord God put him into a deep sleep, removed a rib, and from it created a woman. Since that was undoubtedly a surgical operation, I claim that medicine is the world's oldest profession.'

" 'Wait a moment,' said the architect. 'I must remind you that on the very first day of creation, six days at least before the removal of Adam's rib, the Lord God created heaven and earth out of chaos. Since that has to be considered a structural feat, I maintain that architecture takes pride of place.'

" 'Ah, yes,' purred the lawyer, 'but who do you think created the chaos?' "

And my heart was gladdened when the roar of laughter I got bore promise of being (and indeed turned out to be in the end) the loudest and most prolonged of the entire evening.—And Mr. Nizer was laughing, too, I was relieved to see.

The story has a point right here, too.

In the previous chapter, I spoke about ozone. In our daily life, we encounter ozone (made up of three atoms of oxygen per molecule) because it is formed out of the ordinary two-atom oxygen molecule that is so common in the atmosphere.

But what do you suppose created the ordinary oxygen?

No, not a lawyer—

An atmosphere containing as much free oxygen as ours does is thermodynamically unstable. That means that, left to itself, the free oxygen would gradually disappear. For one thing, it would slowly react with the nitrogen and water vapor in the air and produce nitric acid.

This would happen very slowly, to be sure, but Earth

121

has been in existence for 4.6 billion years. All the oxygen would have combined by now, especially since the energy of the lightning bolt hastens the reaction and produces perceptible amounts of nitric acid, which serve the purpose of helping to renew the dry land's supply of fertilizing nitrates.

If all the oxygen were combined with nitrogen and the resulting nitric acid were to end in the ocean (as it would), then the ocean would be sufficiently acid to make life as we know it impossible.

Well, why hasn't the ocean turned acid long ago? Or if not, why is it not slowly turning acid today?—The small quantities of nitric acid that form nitrates in the soil and ocean are taken up by the living organisms on land and sea and eventually they come out in the form of nitrogen, oxygen, and water again.

The nitrogen and oxygen roll downhill, so to speak, in forming nitric acid, while living organisms kick the nitric acid back uphill as fast as it is formed. Living organisms do this at the expense of energy they gain from chemicals within their tissues, chemicals that were formed originally, in one way or another, by the use of solar energy. It is therefore the energy of the Sun, by way of living organisms, that keeps the oxygen of our atmosphere in its free state and makes animal life, our own included, possible.

This sounds like arguing in a circle. Is life possible only because of something life does? In that case, how did life get started?

The circle isn't really closed, however. It is animal life that can't exist without free oxygen. Nor can any form of animal life maintain an oxygen atmosphere. It is plant life that maintains the oxygen atmosphere and that can, in a pinch, do without free oxygen. Animal life is parasitic on plant life and cannot exist (in the form known to us on Earth) in the absence of plant life.

But then there was a time on Earth when plant life didn't exist either; when no life at all existed. Free oxygen did not then exist in the atmosphere; it couldn't have. Did this mean that the oxygen existed in combination with nitrogen and that Earth had an ocean that was dilute nitric acid? The answer is no, for in that case it seems doubtful that life, as we know it, would have developed.

If oxygen and nitrogen were not combined with each other, they must each have been combined with something else. The only possible something else is hydrogen, which is in vast oversupply in the Universe, which makes up the bulk of the two largest bodies of the Solar System (the Sun and Jupiter), and in which the Earth must have been far richer in primordial days than it is now.

Oxygen combined with hydrogen is water (H_2O), and nitrogen combined with hydrogen is ammonia (NH_3). In addition, the common element carbon can combine with hydrogen to form methane (CH_4). The primordial atmosphere (A-I) could have been made up of ammonia, methane, water vapor, and even some quantities of hydrogen itself. Such a hydrogen-rich atmosphere is called a "reducing atmosphere" for reasons that are buried deep in the history of chemistry and that need not concern us. The present oxygen-rich atmosphere is an "oxidizing atmosphere."

Therefore, when considering the origin of life, it is necessary to imagine processes that would go on in a reducing atmosphere.

If a sample of reducing atmosphere and ocean is left to itself, nothing happens. The various compounds water, ammonia, methane, and hydrogen are a thermodynamically stable mixture, which means the molecules won't alter into anything else unless there is energy present to kick them uphill.

On the primordial Earth, however, there *was* energy. There was the heat of volcanic action, the heat and ionizing power of lightning, the intense radiation of radioactive atoms, and the steady radiation of the Sun. All these energy sources were more intense on the primordial Earth, in all likelihood, than they are today.

In 1952, the American chemist Stanley Lloyd Miller began with a small sample of a mixture similar to the primordial atmosphere, used electric sparks as his energy source, and in the course of a week, found that the simple molecules had combined to form somewhat more complicated molecules, including a couple of the amino acids that form the building blocks out of which those essential life molecules, the proteins, are formed. Since then, further experiments in this direction have made it quite clear that out of the reducing atmosphere plus ocean plus energy a steady series of changes in the direction of life must have taken place.

Can we say which particular source of energy on the primordial Earth was most responsible for life-formation? Consider that, of all the forms, solar radiation is steadiest and most pervasive and it seems logical to give it the lion's share of credit for our presence here today. In particular, we might thank the especially energetic component of sunlight, its ultraviolet radiation. Indeed, experiments have specifically shown that ultraviolet light is energetic enough to interact with the chemicals of the primordial atmosphere and set them off on their march toward life. (Ordinary visible light is *not* energetic enough.)

It seems reasonable, further, to suppose that life began in the surface of the ocean. The ocean is made up of a collection of water molecules and carries many other useful molecules in solution, notably ammonia. Ammonia is so soluble in water that by far the largest portion of it would be in the ocean rather than in the atmosphere. Methane and hydrogen are only slightly soluble in water, but would be in plentiful contact with it at the surface.

The "dry" land is actually moist because of tidal actions, rain, and so on, so it isn't inconceivable that chemicals moving in the direction of life might, to a much lesser extent, form in the soil, but, as I shall explain, they wouldn't get very far.

Ultraviolet light has a rather hammering effect. It can slam small molecules together and make larger molecules out of them, true. But can we then suppose that,

124

as the molecules grow larger and larger under the influence of the ultraviolet, they will eventually become large enough and complex enough to possess the beginnings of life?

Unfortunately, as molecules grow larger, they tend to grow more rickety, and the hammer of ultraviolet is likely to knock them apart again. The influence of ultraviolet, then, may set the primordial molecules to combining in the direction of life, but it won't let them get very far in that direction.

On land, there is no escaping the ultraviolet, so that even if complicated molecules form out of the simple primordial ones, they aren't ever likely to become nearly complicated enough for even the most primitive imaginable life. Life, therefore, cannot start on land.

In the ocean, it is different. Compounds formed on the surface by the action of ultraviolet can, through random motion, sink to a lower level where the ultraviolet cannot penetrate and there they may survive. Indeed, there may be levels where what ultraviolet penetrates can supply energy for combination but not for breakdown.

It would seem then, that in the primordial ocean one would find gradually more complicated molecules as one probed from the very surface downward. The first cases of protoliving substances might have formed some centimeters or decimeters below the surface of the water.

Such life forms may have formed in the first billion years of Earth's existence, and for eons thereafter the situation might have been something like this—

In the topmost layer of the ocean were moderately complex molecules, formed by the energy of the solar ultraviolet, that served as food for the still more complex life molecules below. Some of the food molecules might drift downward and be consumed. More important, on cloudy days or, particularly, at night, the life molecules could somehow drift upward and feed voraciously until the Sun came out, when they would sink again.

We don't know to what level of complexity life forms

may have developed in this period. The only traces of life that we can find that date back a billion years or more seem to have arisen from tiny one-celled creatures and nothing more. This is perhaps not surprising. It doesn't take much energy to change ammonia, methane, and water into food chemicals and, conversely, it doesn't release much energy to break them down again.

Primordial life did not have much energy at its command and it could live and evolve only slowly.

All might have continued to be so to this day if the A-I atmosphere had remain unchanged—but it didn't.

For one thing, it lost its hydrogen. Any hydrogen that the primordial Earth may have had in its atmosphere was quickly lost to outer space, since Earth's gravity could not hold its small and quickly moving molecules.

Then, too, the ultraviolet light of the Sun, at its full strength in the upper atmosphere, can hammer apart even simple molecules. The water molecule, particularly, can be broken down to hydrogen and oxygen by the action of ultraviolet. This is called "photolysis."

The photolysis of water takes place only high in the atmosphere for the most part. Few water molecules are found so high and the process is slow—but again Earth is long-lived and has time.

The hydrogen that is produced by photolysis is lost to space, but the heavier, less nimble oxygen atoms remain behind. In the presence of free oxygen, however, methane and ammonia are no longer thermodynamically stable. The carbon and hydrogen atoms making up the methane molecule tend to combine with oxygen atoms to form carbon dioxide (CO_2) and water respectively. The hydrogen atoms of the ammonia molecules combine with oxygen to form water, leaving behind the nitrogen atoms, which combine to form two-atom nitrogen molecules (N_2). The nitrogen would also combine with oxygen, but so much more slowly that the carbon and hydrogen atoms get all the oxygen.

The net result is that the methane/ammonia/hydrogen/water-vapor atmosphere (A-I) is slowly converted, through photolysis, to a carbon-dioxide/nitrogen/water-vapor atmosphere (A-II).

To kick the molecules of A-II uphill to the level of the food molecules took more energy than was the case starting with the molecules of A-I. For that reason the rate of food production declined and as A-I slowly changed to A-II, a kind of famine spread over the face of the ocean.

The type of organisms that had developed in A-I and that lived on the breakdown of food molecules to ammonia and methane, and that made do with the small quantity of energy made available in the process, must gradually have decreased in numbers in the face of the spreading famine.[*]

Once A-I had completely turned to A-II, it might seem that the food situation had hit bottom for A-I organisms, but not so. Things grew worse still because of photolysis.

Even after the atmosphere had become completely A-II, photolysis continued to take place, water molecules to break down, hydrogen atoms to escape, and oxygen atoms to remain behind. But now, the oxygen atoms had nothing to combine with but each other (or, very slowly, with nitrogen). Ordinarily, they would form the two-atom oxygen molecule, but in the upper atmosphere they can, under certain conditions, be kicked further uphill by the energy of ultraviolet light and form three-atom ozone molecules.

Ozone molecules are opaque to almost all the ultra-violet range. As more ozone forms, less and less ultra-violet manages to penetrate past it. Thus not only did the A-II atmosphere possess molecules more difficult to change into food, but it began to allow less and less ultraviolet through to make any change at all.

With less and less ultraviolet available, the rate of

---

* They never entirely disappeared, to be sure, for there are still organisms alive today that live on types of chemical reactions other than those of most life forms. They are very likely descendants, fundamentally unchanged, of A-I life forms.

photolysis (carried on at atmospheric heights well below the regions where ozone is formed) would dwindle as well. This would mean that Atmosphere II would stabilize and that further change would become less and less likely, but only after the ultraviolet supply at the ocean surface had been virtually shut off.

At the present time, ozone is concentrated between heights of twenty-five to sixty-five kilometers above Earth's surface (the "ozonosphere"), but even there only one molecule out of a hundred thousand (of an atmosphere that is excessively thin at such heights) is ozone.

Even though the ozone molecules are exceedingly rare by ordinary standards, they suffice to shut out almost all the ultraviolet and leave very little to reach Earth's surface. (Enough to sunburn light-skinned people such as myself, to be sure, so I'm intelligent enough to stay out of the sun.)

Life on Earth would have had to dwindle to a very low level, supported by the subsidiary energy sources of lightning, radioactivity, and volcanic heat, and have continued so indefinitely, but for an unexpected happening.

Somehow (we don't know the details) and at some time (we don't know exactly when) the most important evolutionary development, next to the beginning of life itself, came about. A chlorophyll-like molecule must have been developed, together with a primitive enzyme system, capable of catalyzing the combination of carbon dioxide and water to form food molecules. This was the beginning of "photosynthesis."

The development of photosynthesis by organisms adapted to A-II meant the following—

1) Until then, ultraviolet light was the motivating force for the production of food, but photosynthesis made use of the less energetic wavelengths of visible light. Since visible light is more copious in solar radiation than ultraviolet is, it could be the source of a potentially much larger supply of food.

2) Since photosynthesis takes place amid the very molecules of the life form, the food is formed there and does not have to be hunted for in the ocean generally. This must have meant that cells could become larger and more complex.

3) Since visible light is *not* blocked by ozone, the A-II photosynthetic organisms were unaffected by the slow closing of the ozone curtain and could flourish even while the A-I organisms faded.

4) In converting methane, ammonia, and water into food chemicals, the general atomic makeup remains largely unchanged and there is little "waste" left over. In using water and carbon dioxide as the source of food, however, we begin with molecules containing more oxygen atoms than is required for the food. Those oxygen atoms must be discarded as "waste" and are dumped into the atmosphere.

The existence of photosynthesis, then, hastened the rate at which free oxygen was poured into the atmosphere. Indeed, photosynthesis produced free oxygen at a rate far, far beyond that of photolysis. The ozone curtain began to close at a precipitously greater and greater rate, so that A-II life forms, by means of the new chemistry they had evolved, actually hastened, very markedly, the demise of the A-I life forms. Without moving from the spot and without any obvious aggression, they won a silent victory of planetary scope, one scarcely rivaled in extent since.

5) Photosynthetic life forms flourished so mightily that they consumed the carbon dioxide of the atmosphere, incorporating the carbon into their own tissues and filling the air with oxygen instead. In this way, through the action of life, the nitrogen/carbon-dioxide A-II was changed over to the nitrogen/oxygen A-III of today.

The concentration of carbon dioxide in the atmosphere today is only 0.035 per cent as compared to oxygen's 21 per cent. In fact, it would seem to be useful to the plant world, generally, to have as parasites on them life forms that consume oxygen and produce carbon dioxide. These would serve to increase, at least by a little, the carbon dioxide of the air. Thus A-II life forms were differentiated into plants and animals, whereas A-I life forms may never have advanced beyond the bacterial stage.

6) A-II life forms developed enzyme systems capable of handling the very active oxygen molecules. A-I life forms apparently did not do so. Free oxygen was an active life-destroying poison to them and in this way, too, A-II life forms accelerated their silent victory.

7) Since the energy required to change carbon dioxide and water to food is extraordinarily high by A-I standards, the reconversion of food to carbon dioxide and water liberates an extraordinarily high level of energy. This means that A-II life forms had much more energy at their command than did the A-I life forms. This was especially true of the A-II animals, who could make use of the food supply of many plants at once.

When did A-II organisms start pouring oxygen into the atmosphere?

We can't tell. Photosynthesis may have developed quite early on but may have remained very inefficient for many millions of years and its production of oxygen may have been very, very slow. A-II organisms may have barely struggled along in the shadow of more successful A-I organisms for a long time.

When did photosynthesis become efficient enough and the oxygen supply of the atmosphere high enough to mark the completion of the silent victory of A-II?

My guess is about 700 million years ago. There must have been a time when photosynthetic efficiency rose so high that there was an explosion of evolutionary energy, and it is about 600 million years ago, quite suddenly, that complex life forms began to be present in such quantities as to begin to leave copious fossil traces. By then we began to have A-III organisms, well above the A-I and A-II organisms in complexity.

And when was the changeover to A-III complete?

My guess is 400 million years ago. By that time, although life had existed for over three billion years, the dry land had not yet been colonized. In an earlier article,† I suggested that the dry land was not colonized until after Earth had captured the Moon and had been subjected to tidal effects. That might be so, but I must admit a more likely explanation of the delay in the colonization of the land now occurs to me, and one that I have not seen advanced elsewhere.

After all, while ultraviolet bathes the surface of the Earth, any attempt on the part of life to emerge on dry land would mean steady exposure to ultraviolet without the easy escape hatch equivalent to the oceanic device of sinking a bit deeper in the water. It was only after the ozone curtain was closed that the dry land became safe for life and it was 400 million years ago that life began to scramble ashore.

Now, then, what if anything happens to that dreadfully thin and, perhaps, fragile ozone layer?

The changes made possible by the closing of the ozone curtain would be reversed. Once again, the Sun's ultraviolet would flood the face of the Earth so that the planet's land surface together with the topmost skin of the ocean would become as inimical to life as it had been more than 400 million years ago. What's more, the photolysis of water molecules would begin again.

Need we panic, however? After all, even if photolysis begins again, it would take billions of years to exhaust the ocean. And land life today is not what it was 400 million years ago. Animals have skin, scales, hair, feathers, all of which block off ultraviolet and prevent instant damage to internal organs.

Then, too, advanced animals can seek the shade and the most advanced animal, *Homo sapiens,* can use umbrellas, build glass barriers, move further poleward and so on. Even the complete opening of the ozone curtain might not serve to damage, seriously, the advanced life

† "Triumph of the Moon" in *The Tragedy of the Moon* (Doubleday, 1973).

forms on Earth, or to do more than inconvenience, perhaps, humanity in general.

It would increase the incidence of human skin cancer, especially among the light-skinned, if we didn't take precautions, and it might accelerate the mutation rate, particularly among plants, with unpredictable results, but what else?

Well, not all land life forms are advanced beyond their beginnings. There are still protozoa, algae, bacteria, and viruses who have no particular protection against ultraviolet and have no behavioral patterns that would help them escape it. If the ozone curtain is ripped open, it may be that land micro-organisms will be seriously depleted—and we don't know what that will do to the rest of the ecological structure.

How will the death of the micro-organisms affect the nature of the soil, the growth of crops, the life of animals, including man. We don't know how, but it seems to me it can scarcely do us any good, and it may just possibly represent a colossal disaster.

Is there, then, anything which endangers an ozone curtain that has remained shut for at least 400 million years. Concerning that, I will have something to say in the next chapter.

10

Change of Air

I seem to nonplus even my nearest and dearest on occasion. You would think they would know my pecularities—

My wife, Janet, and I were crossing West Virginia by auto about six weeks ago and we stopped off at a lodge that was located rather high on the side of a mountain just about in the middle of the state.

After dinner, we wandered out on the grounds and managed to make our way to a rocky ledge (well-fenced) that overlooked the gorge through which a river wound its way. Janet, who is a great admirer of natural vistas, was caught up in its beauty; while I, who am acrophobic and don't like to look down, and who prefer to have my views of nature by color photograph, stood beside her a little uneasily.

The cloudless sky was still bright, but the twilight was deepening; the vista was absolutely bursting with green; the river was silver below; and around the bend of a mountain there slowly came a long freight train dragged by four locomotives. It crawled its way precariously along the narrow space between mountain and river, with its busy chug-chug far enough away to sound like the panting of a giant anaconda.

After a long while, Janet said, in an awed whisper, "Isn't this amazing?"

"You bet," I said, briskly. "One hundred sixty-six cars! Longest freight train I ever saw!"

I disregarded her threat to push me off the ledge. I knew that, despite everything, she was too fond of me to make the attempt.

The trouble is, you see, that some people have a prejudice against counting and measuring and weighing. They just want to look at things qualitatively. Yet sometimes careful measuring of tiny things may prove a matter of life and death for you and me and all of us, as I will demonstrate before I am through with this article.

In the two preceding chapters, I discussed ozone and talked about the development of the ozonosphere. Now we will move into organic chemistry, and eventually make a connection.

The characteristic molecules of living matter are made up of chains and rings of carbon atoms. Almost every carbon atom is attached to one or two other carbon atoms and to one or two hydrogen atoms in addition. Occasionally, a carbon atom is attached to an oxygen atom or to a nitrogen atom, and very occasionally to a sulfur atom.

In nature, that about exhausts the kinds of atoms to which the carbon atom is attached. In the early days of organic chemistry, it was thought that other types of atoms could not be attached to carbon atoms. In particular, it seemed that atoms of the newly discovered element chlorine, being radically different in properties from hydrogen, could not replace hydrogen atoms along the carbon chain.

This theory was smashed in the most direct possible way. A molecule was formed in which the carbon-chlorine connection existed. In 1834, a French chemist, Jean Baptiste André Dumas (no relation to the novelist), formed "chloroform." The chloroform molecule contains a single carbon atom attached to one hydrogen atom and three chlorine atoms ($CHCl_3$).

Chloroform didn't remain an exotic laboratory product for long. The concept of chemical anesthesia arose not long after and a Scottish physician, James Young Simpson, began using chloroform as an anesthetic in 1846. In 1853, he used it on Queen Victoria in childbirth and it became a household word. In fact, such is

134

the general stupidity of *Homo asinus* that people began to hold "chloroform parties." They would sit around bowls of chloroform, inhaling the fumes, till they fell over unconscious. What they got out of it aside from ruined livers, I can't imagine.

The danger of chloroform was such, in fact, that it was quickly outpaced by diethyl ether as an anesthetic. In fact, chloroform is used as an anesthetic today chiefly in books and movies of the sillier sort.

After the discovery of chloroform, all sorts of other "organic chlorides" were formed. There were even molecules in which carbon atoms were attached *only* to chlorine atoms, and which might therefore be called "chlorocarbons" in analogy to "hydrocarbons," which have molecules made up of carbon and hydrogen atoms only.

The simplest of the chlorocarbons is "carbon tetrachloride," with a molecule consisting of one carbon atom attached to four chlorine atoms (CCl_4). Another is "tetrachloroethylene," in whose molecule carbon atoms, attached to each other by a double bond, are hooked up to four chlorine atoms ($CCl_2{=}CCl_2$).

Both chlorocarbons and hydrocarbons readily dissolve molecules of fat and grease, but whereas the hydrocarbon molecules are easily inflammable and present a definite fire hazard, the chlorocarbons are *not* inflammable. Indeed, carbon tetrachloride can be used in fire extinguishers. For that reason, chlorocarbons, particularly tetrachloroethylene, are used as dry-cleaners—though it is best not to breathe the vapors, which are quite poisonous.

By the time chloroform had been produced, two elements had been discovered that resembled chlorine in chemical properties. These were bromine and iodine, which, with chlorine, are grouped together as "halogens" (from Greek words meaning "salt-formers"). In combination with sodium, each of these forms saltlike compounds. Chlorine, indeed, with sodium, forms sodium chloride, which is *the* "table salt" we use at our meals.

The carbon atom, it turned out, would combine with

135

any of the halogens. Compounds analogous to chloroform are bromoform ($CHBr_3$) and iodoform (CHI_3).

Because iodoform had some disinfectant properties and wasn't too damaging to tissue, it came to be used in dressing wounds and for a short while doctors and hospitals smelled of iodoform. They still do so, even today, in books written by writers who get their clichés from other books.

You also have carbon tetrabromide (CBr_4) and carbon tetraiodide (CI_4), which are examples of "bromocarbons" and "iodocarbons" respectively.

There is a limit to the number of halogen atoms you can pack in around the carbon chains and rings. Hydrogen atoms are the smallest there are so that they can attach themselves to any available site on any carbon atom, no matter where in a chain or ring that carbon atom is located. There is always enough room.

Chlorine atoms, however, are considerably larger than hydrogen atoms, bromine atoms are even larger, and iodine atoms are larger still. Too many of them attached to too many carbon atoms in a single molecule tend to get in each other's way. It is therefore difficult to get any very large chlorocarbons and almost impossible to get large bromocarbons or iodocarbons.

But there is a fourth halogen. For decades after the discovery of the first three, chemists were certain there was a fourth, lighter than the others, one which they could not isolate because its atoms clung so tightly to other atoms. They called the new halogen "fluorine" even before it was isolated and you will find the tale of that isolation in my essay "Welcome, Stranger."*

Everyone was certain that once fluorine had been isolated and its chemical properties studied, it would be found to attach itself to carbon atoms to form "organic fluorides." It was also certain that there would be cases where carbon atoms were attached *only* to fluorine atoms to form "fluorocarbons."

* In *Of Time and Space and Other Things* (Doubleday, 1965).

The French chemist Ferdinand Frédéric Henri Moissan, who finally isolated fluorine in 1886, tested the matter at once. He found that fluorine and hydrocarbons mixed might indeed produce fluorocarbons as the fluorine atoms replaced the hydrogen atom on the carbon chain—but he couldn't tell because the mixture exploded at once, and disastrously, and there was no way of analyzing for fluorocarbons among the shattered equipment.

When he switched to plain carbon in place of the hydrocarbons (carbon reacts much more slowly than hydrocarbons do) he got no improvement. The fluorine reacted with powdered carbon explosively.

In 1905, Moissan had another idea. He decided to mix fluorine and methane (CH_4) but to do it at a very low temperature—at liquid air temperatures in fact. At $-185°$ C, with methane frozen to a solid and fluorine a liquid nearly at its freezing point, he mixed the two and got one more disastrous explosion.

It was not till forty years had passed after the isolation of fluorine that progress in organic fluorides was made. In 1926, two French chemists, P. Lebeau and A. Damiens,† managed to burn carbon in fluorine gently enough to be able to study the products formed. They found that carbon tetrafluoride (CF_4) was formed. It was the first organic fluoride (and fluorocarbon) to be obtained in pure form.

In the early 1930s, two two-carbon fluorocarbons were obtained pure. These were hexafluoroethane (CF_3CF_3) and tetrafluoroethylene ($CF_2=CF_2$).

The study of fluorocarbons wasn't going to get much farther, however, until some way was discovered of taming fluorine. Low temperatures weren't good enough, but perhaps some substance could be added to the reacting mixture; some substance that would act as a

† I like to give the names of scientists in full, even when they have three middle names, since they so rarely get the exposure that, in our society, basketball players and country-music singers get as a matter of routine. The trouble is that I can't always locate the full name. If any of my Gentle Readers should ever recognize a friend under the initials, please let me know.

catalyst or an intermediary in the reaction, allowing it to proceed in more orderly fashion.

In 1934, for instance, the German chemist Karl Hermann Heinrich Philipp Fredenhagen found that if fluorine were introduced through a copper-mesh screen to the hydrocarbon it was to react with, the reaction would be more moderate.

Then, in 1937, the American chemist Joseph H. Simons found that if powdered carbon were mixed with a small quantity of a mercury compound, the carbon would burn in fluorine more quietly and produce a variety of fluorocarbons in the process. Simons was, in fact, able to produce and study fluorocarbons with molecules containing up to seven carbon atoms.

This was the first real indication that fluorocarbons could be formed much more easily than any of the other halocarbons and would involve long carbon chains. At that, this was not surprising. The fluorine atom is smaller than those of the other halogens, and when it is attached to a carbon atom, it takes up less room than does any other atom but hydrogen. There is, actually, room for the fluorine atom to hook onto any carbon atom, whatever its position in a chain or ring and adjacent fluorine atoms are small enough not to get in each other's way. A fluorocarbon can be formed that is analogous to any hydrocarbon.

Simons was particularly impressed with the stability of the fluorocarbons and with their inertness. The fluorine atom holds on to a carbon atom more tightly than a hydrogen atom does, to begin with. What's more, as additional fluorine atoms are attached to the carbon chain, they seem to reinforce each other and their bonds grow tighter still. By the time all the hydrogen atoms are replaced, the bonds holding the carbon and fluorine atoms together are so tight that almost nothing will budge them. Fluorocarbons will not burn, dissolve in water, or react with almost anything.

Meanwhile, as the 1930s came to a close, the American chemist Harold Clayton Urey was working with uranium and trying to isolate, or at least concentrate,

138

uranium-235 (and you know why). If he could discover some gaseous compound of uranium, the molecules containing U-235 would move a little faster than those containing the heavier U-238 and he might manage to separate the two in that way.

The only uranium compound that would become gaseous at reasonable temperatures was "uranium hexafluoride" (UF_6), but this compound had a tendency to react with the substances used to seal and lubricate the joints of the system within which the separation was to take place.

Joseph Simons heard of this and it occurred to him that a liquid fluorocarbon would be stable enough not to be attacked by uranium hexafluoride and it could therefore be used as a lubricant. He had perhaps some forty or fifty drops of what he thought might be suitable material, and he sent nearly all of it to Urey in 1940. It worked and was referred to thereafter, in elliptical fashion, as "Joe's stuff."

From then, on, there came a big push to form more fluorocarbons. All sorts of elaborate schemes were used, such as using catalysts *plus* low temperatures, using metal fluorides as a source of fluorine rather than fluorine itself, using organic halides rather than hydrocarbons to react with fluorine, using hydrogen fluoride and an electric current, and so on.

What it amounted to was that by the end of the war, fluorocarbons were common. In fact, one could form long chains of carbon atoms, with fluorine atoms attached at every point—"fluorocarbon resins." To do this you begin with tetrafluoroethylene (CF_2—CF_2), which has a double bond in the middle. One of those bonds can open up and neighboring molecules can hook together by means of that bond and thus form a long chain. Dupont called a substance with a long-chain molecule of this type "Teflon," and it is most familiar to us as a lining for frying pans. It is stable enough not to be affected by the heat of frying, and inert enough not to stick to any of the food, so that it is easy to clean.

At Dupont, compounds were formed consisting of molecules in which carbon atoms were attached to both chlorine and fluorine atoms and nothing else ("fluorochlorocarbons"). It was found, rather unexpectedly, that the presence of the fluorine atoms served to tighten the chlorine-carbon bond so that these mixed organic halides were as stable and inert as fluorocarbons themselves—and cheaper because not as much fluorine had to be used. Dupont named one of this new class of substances "Freon."

Freon revolutionized the technique of refrigeration.

The use of ice to cool household perishables had been replaced by electric or gas refrigerators, which made use of either a gas that is easily liquefied or a liquid that is easily vaporized.

In either case, the liquid is pumped through pipes in a closed container, the refrigerator, and is allowed to vaporize. The process of vaporization requires an input of heat, which the liquid absorbs from the substances inside the refrigerator. These substances therefore cool down. The gas then emerges from the container and is condensed into a liquid again, giving up the heat it had absorbed when it had vaporized. The heat is carried away by a coolant of some sort (air or water) and the cooled liquid goes through the refrigerator to vaporize again. Heat is thus steadily pumped out of the refrigerator into the open air.

Before World War II, the most common liquid used for refrigeration was ammonia. To a lesser extent, sulfur dioxide or some simple organic chloride was used. They worked very well as refrigerants, but they tended to corrode the pipes, and if leaks developed there was the distressing fact that they had choking odors and were poisonous. There was a limit, therefore, to how thoroughly refrigeration could be adapted for home use.

But then came the various Freons. Some of them were easily vaporized liquids, and some were easily liquefied gases. They were inert and wouldn't react with anything they came in contact with. If a leak did develop for any reason and Freon got into the air, there was neither smell nor damage. Freon could be inhaled with no harm to the body at all. Indeed, one of the early workers with Freon, Thomas Midgley, Jr., demonstrated its harmlessness by taking in a deep lungful and letting it trickle out over a lighted candle. The candle went out but Midgley was unharmed. (Of course, if Midgley had kept on breathing pure Freon he would have suffocated for lack of oxygen—but not out of any direct harm the Freon had done.)

The most common Freon varieties used for refrigeration are Freon-11 (CCl_3F) and Freon-12 (CCl_2F_2), the latter particularly. It is the convenience and safety of the Freons that has resulted in the almost universal use of air conditioning today.

Then, too, there is the matter of spray cans. It is very convenient to be able to apply certain materials in the form of a spray. One way of doing this is to force a liquid through a small hole. The liquid is divided into tiny particles suspended in air as a mist and this is called an "aerosol."

Such forcing can be done by muscle power, but that would be tiring. It can be done by gas pressure, say by having a bit of solid carbon dioxide, or liquid carbon dioxide under pressure, evaporate and produce gas pressure inside the can. Such pressure will force out liquids or soft solids as mists or foams. But then you have to have a strong steel can to retain the pressure till you want the use of it.

In the 1950s, it occurred to the Dupont people that a mixture of Freon-11 and Freon 12 could be used for the purpose. The liquid mixture would develop enough pressure to produce the spray, but would build up a far smaller total pressure. It could be safely contained in a thin, light aluminium can. Of course, the Freon emerged with the spray, but it was even less dangerous than carbon dioxide would be.

What made this particularly useful was that at about this time, Robert H. Abplanalp invented a simple plastic-and-metal valve that could be manufactured cheaply

141

and made part of the spray can. The touch of a finger would release the spray and the removal of the finger would stop it.

At once, the spray can came into fashion and Abplanalp became wealthy enough to qualify as a great friend of Richard Nixon. In 1954, 188 million aerosol spray cans were produced in the United States. Twenty years later the annual production had passed the three billion mark, a sixteenfold increase.

But what happens to all the Freon that is produced and used in refrigerators, air-conditioners, and spray cans? In the case of the spray cans, it is obviously discharged into the air. The Freon in refrigerators and air-conditioners is also eventually discharged into the air, since these devices will sooner or later leak or break. Every bit of Freon produced, and it has been produced by the millions of tons, will be discharged into the air.

This is not, in itself, a frightening thing. The air is the common sewer for an incredible number of substances. Volcanoes spew cubic miles of materials into the air. Forest fires are almost as bad. Even so unglamorous an event as the farting of herbivorous animals discharges enough methane into the air to build up a detectable quantity.

But all these naturally produced substances don't stay in the air. Dust particles settle out; gases are washed out by the rain and react with the soil. On the whole, what enters the air leaves the air, and this is true, in general, even for man-made products. Sometimes human activity pushes substances into the air faster than it can be removed, so that the carbon dioxide and dust content of the air is higher than it would be if mankind lived a primitive life. However, if ever mankind did decide to live a primitive life, those impurities in the air would decline soon enough. No natural process, over the short term, permanently changes the air. Nor does any man-made activity.

—Except one.

Freon changes the air. It enters, but it does not leave.

It doesn't settle out; it doesn't wash out; it doesn't react out. It just stays in the air and it has been accumulating for a quarter of a century. In this respect, mankind has produced a permanent and perceptible change in the chemical composition of the air, and it is continuing to intensify this change of air.

Is that bad? No, not if we consider the lower atmosphere only. Freon does us absolutely no harm and although it has been discharged into the atmosphere by the millions of tons, the atmosphere is so voluminous that the concentration of Freon is still minuscule and will remain so for a long time.

Even if the concentration of Freon in the air becomes appreciable and we nevertheless still insist on pouring more of the stuff into the air, there is a natural limit. Eventually, there will be enough Freon in the air to make the atmosphere a practical source of the gas. We can then take the gas out of the atmosphere and pour it back again after using and there will be a steady level which will probably still be harmless to us. (Taking the gas out and putting it back in again will cost energy of course, but so what? Everything costs energy.)

So far that sounds fine but then, in 1973, F. Sherwood Rowland, a chemist at the University of California, began to consider the problem.

There is, after all, one way in which the Freon molecules *can* break down. Ordinary light doesn't affect Freon but ultraviolet light, which is more energetic than ordinary light, is energetic enough to break chlorine atoms away from the Freon molecule.

Even this would not be serious down here. In the first place, thanks to the ozone layer about twenty-five kilometers and more above us, not much ultraviolet light gets to the lower atmosphere. Then, too, even if tiny bits of chlorine are produced, so what! They will dissolve in water, or react with other substances, and never accumulate to an amount that will be in the least harmful to us.

But what if the Freon molecules gradually diffuse upward to the ozonosphere and above? At those heights there is a great deal of energetic ultraviolet light from the unfiltered Sun and perhaps every Freon molecule that makes its way up there will then break down and produce chlorine atoms.

You might still say, so what! Very few molecules will make it up to that height, and the tiny bits of chlorine will just react with something and be consumed.

But that's the trouble! The chlorine atoms *will* react and will do so, among other things, with the ozone molecules of the ozonosphere. The chlorine atom will combine one of the three oxygen atoms of the ozone molecule, forming chlorine oxide (CIO) and leaving the two oxygen atoms of the ordinary oxygen molecule behind.

Ought that to worry us? There is a considerable amout of ozone up there. The ozone is constantly breaking down and being re-formed, and the little bit of extra breakdown produced by an occasional chlorine atom should be an insignificant matter.

But it isn't. The chlorine oxide that is formed will eventually combine with a free oxygen atom that is occasionally formed by the natural breakdown of ozone. The chlorine oxide will give up its oxygen to the oxygen atom so that ordinary oxygen molecules are formed. A free chlorine atom is left behind which can then attack another ozone molecule.

Each chlorine atom, as it combines with an oxygen atom and then gives it up again, can react with ozone over and over and over again. Each chlorine atom can break down not one ozone molecule but perhaps hundreds before anything happens to break the chain.

It would seem then that the amount of Freon drifting up into the ozonosphere can damage it by an amount far out of proportion to what might be expected from its mere quantity. Furthermore, there is a delayed-action effect here, because even if mankind stops using Freon altogether at this very moment, the quantity that has already been discharged into the atmosphere, plus the amount now present in all Freon-using devices, which will eventually be discharged, will continue upward for years to come from now.

But how much damage will there be? Will it be enough to deplete the ozonosphere seriously and to bathe Earth, for the first time in hundreds of millions of years, with markedly high concentrations of solar ultraviolet light, producing the possible damage I mentioned at the end of the last chapter?

We are not sure yet. We don't know the rate at which Freon molecules arrive in the ozonosphere, nor the exact nature of the reactions that go on there. We don't know the details of the natural processes that break down and re-form the ozone.

First reports made things look bad, but since then there has been a retreat from some of the more alarming estimates and at the time of writing there seems a notable decline in panic over the possible disappearance of the ozonosphere.

Ignorance, however, is no substitute for security. The Freon may not do us harm, but it may, and we had better find out ways of knowing what's going on. We had better bend every effort to working out methods for measuring the density of the ozonosphere and we had better keep it under constant monitoring. That kind of delicate and constant measuring (as I indicated at the end of my introduction to this article) could be a matter of life and death.

And, just in case, I do think we ought to moderate the use of Freon. A little inconvenience till we can make sure no harm is being done is surely better than a panic stop to its use ten years from now—when it may be too late.

11

The Wicked Witch is Dead

Good heavens, I've become a household word. I suppose that's inevitable if one lives long enough and writes very prolifically on a wide variety of subjects in a wide variety of outlets. —But I'm not sure I'm comfortable with it.

I'm constantly being quoted, for instance. In a recent three-hour TV special on the role of women, Barbara Walters quoted "the famous science writer Isaac Asimov," and, as usual, I missed it. The news was brought to me by others, and no one remembered *what* she had quoted. What can I have said, I wonder?

Then, a week ago, I came dashing out of a TV studio to grab a taxi for my next appointment and the driver was curious enough to ask what I had been doing there. I explained that I was a writer who had just been interviewed, and the driver admitted that he himself was trying to be a writer but had had no luck so far.

"Well," I said, trying to be consoling, "don't feel bad. Writing is a mug's game, anyway. Hardly any writer can make a living out of it."

"Isaac Asimov does," said the driver, broodingly—and left me speechless.

My favorite story of this sort, however, took place about a year ago, when Janet and I had been at the theater for the evening. We were back by about 11:40 P.M. and I turned on WQXR in order to make sure I got the midnight news so that I might find out if there were any late word on Watergate. (I was an insatiable Watergate fan.)

A voice came out of the radio and, as I listened absently, I began to find the statements familiar. I called out, in annoyance, "Hey, Janet, there's a joker here who's spouting my ideas, word for word."

Janet came over, listened for a moment, and said, patiently, "'It's *you*, dear."

And so it was. I had been interviewed and taped a month before and had been just in time, inadvertently, to catch myself. It's just so difficult to recognize one's own voice when it isn't resonating in one's own skull cavities.

It's just going to get worse, too, since I have every intention of continuing to express my views on every conceivable subject—the more controversial, the better.

As on the matter of witches, for instance.

Witchcraft, apparently involves the use of supernatural methods for controlling the energies of the Universe toward, supposedly, some evil or destructive end.

Those who admitted the existence of witchcraft had no doubt that there were indeed supernatural powers who could be controlled by human beings using appropriate methods. Indeed, what else is the *popular* view of "true religion" in the Western tradition but an attempt to use supernatural methods for controlling the energies of the Universe toward, supposedly, some good or constructive end?

The attempt to control supernatural powers is known as "magic" (though the word has been weakened these days to signify mere prestidigitation or illusion). The word is derived from *magu*, the name given by the ancient Persians to their Zoroastrian priests; and, indeed, magic is the priestly function.

We don't use the word in connection with our own religion but, really, doesn't the use of prayer represent an example of magic? By using appropriate words and, sometimes, music; by using sonorous foreign languages, incense, and other impressive adjuncts, we hope to persuade, cajole, or irritate a particular supernatural power into suspending the natural laws of the Universe just long enough to arrange things to suit our immediate personal needs—to make it rain, for instance, when the supernatural plan may be for a long drought.

Well, then, if religion uses the supernatural to good ends; and if witchcraft uses it to evil ends; a lot depends on how we define good and evil.

It is not surprising that, generally, the final decision comes to this: that what *we* do is good, and what *they* do is evil.

The Bible sanctions magical practices for foretelling the future or, to use an equivalent phrase, for ascertaining the will of God. The Urim and Thummim, for instance, seem to have been lots kept in the vestments of the High Priest and these could be cast, with appropriate rites, in such a way as to yield a marking that could be interpreted as a direct indication of the Divine will.

Attempts to use similar devices by any methods or rites not sanctified by the dominant religion (whatever it happens to be) are, of course, roundly condemned as dealing with demons, who are usually known in our language as "familiar spirits."—The word "familiar" refers to a spirit bound to the service of a single individual and hence part of the family, so to speak.

Those who possess such a familiar spirit are "wizards" or "warlocks." Since, in Anglo-Saxon, a wizard is a "wise man" and a warlock is a "deceiver," I suspect the two terms were used, originally, for "our" priests and "their" priests, respectively. A "sorcerer" is from a Latin word for "lot" and is therefore someone who ascertains the future by the use of lots such as the Urim and Thummim.

A feminine practitioner of these arts is a "sorceress" or a "witch." The word "witch" is from the Anglo-Saxon *wicca,* which may be related to "victim," so that a witch is one who presided over sacrifices—"their" wicked sacrifices, of course, not "our" good ones. A female conductor of sacrifices according to "our" rites is, of course, a "priestess."

When a land comes under a strong rule, and the ruler is committed to one particular ritual, then the state rite becomes "religion" and all other rites become "witchcraft" and must be suppressed. This was the case in Israel, for instance, when Saul was king. "And Saul had put away those that had familiar spirits, and the wizards, out of the land" (1 Samuel 28:3).

But when Saul was faced with a crisis and turned to the Lord for guidance, he failed. "And when Saul enquired of the Lord, the Lord answered him not, neither by dreams, nor by Urim, nor by prophets" (1 Samuel 28:6).

Saul was forced by the gravity of the crisis to turn to other methods. "Then said Saul unto his servants, Seek me a woman that hath a familiar spirit, that I may go to her, and enquire of her. And his servants said to him, Behold, there is a woman that hath a familiar spirit at En-dor" (1 Samuel 28:7).

This woman is not specifically called a witch in the Bible, but she is commonly known today as "the witch of Endor." It is for that reason that, in the TV program "Bewitched," Agnes Morehead played a witch whose name was "Endora."

The witch of Endor managed to bring up the spirit of the dead prophet Samuel, who was questioned by Saul and who predicted disaster.

This tale is biblical justification, if you care to use it, for the real existence of witches, for the effectiveness of magic and of familiar spirits, for the ability to raise the dead and predict the future.

If we interpret the Bible literally, we cannot think that witchcraft is a delusion. No, it is a competing religion making use of powerful supernatural forces, and, because it is competing, it is evil by definition.

Any religion which is certain it has the "truth" finds it very difficult to tolerate the existence of another religion and, in the Bible, such toleration is not suggested.

149

In Leviticus 20:27, there is the following: "A man also or woman that hath a familiar spirit, or that is a wizard, shall surely be put to death: they shall stone them with stones: their blood shall be upon them."

In Exodus 22:18, there is the briefer: "Thou shalt not suffer a witch to live."

These verses may be the first clearly expressed examples of religious intolerance in the world. They, and the spirit in which they were written, have been used to justify the spilling of rivers of blood and the infliction of countless horrors.

The attitude of intolerance toward competing religions continues, of course, into the New Testament, where the Gospels are haunted by tales of demonic possession. There are references to exorcists capable of controlling evil spirits through their magical rites (see Acts 19:13–16).

During the Middle Ages, there remained remnants of pre-Christian religion in Europe, old peasant beliefs that harked back to pagan days. The old paganism had not died but had been driven, in more or less altered and debased form, underground. There were still rites, secretly practiced, involving a horned goat-god and actions designed to promote fertility in the customary fashion of primitive agricultural magic.

These practices were naturally stigmatized as witchcraft, but the Church, secure in its power, and concerned with more dangerous manifestations of competition, in the form of Moslems without and various Christian heresies within, did little, for many centuries, to combat these beliefs except through verbal denunciation.

The matter changed as strife grew within the Church, culminating in the Protestant schism that began in 1517. More and more there came to be the feeling of dangerous competition, of deep insecurity on both sides. Religious intolerance heightened and strengthened so that for a century Catholics and Protestants continually fought each other by battle when the two sides were roughly equal in strength, or by slaughter when one side or the other was overwhelmingly powerful.

The constant exercise of intolerance heightened both the feeling of self-righteousness in the pious and their conviction of the infinite evil of all those who denied the "true religion." Both Catholics and Protestants amused themselves, therefore, in the intervals when they were not too busily engaged in killing each other, in hounding down all those who were accused (the accusation was usually enough) of dealing with the devil. From 1500 to 1750, the mania continued, with a faint echo in Salem, Massachusetts, in 1692.

And what brought an end to this? That which brought an end to all magic, whether the holy rites of "our" religion or the blasphemous witchcraft of "their" religion.

Science, making use of the natural laws of the Universe, and doing so in a demonstrably workable fashion, became the approved method of forcing man's desires upon the world.*

If the president should fall ill, I am sure that the nation's houses of worship would organize prayers on his behalf as a matter of social reflex. I doubt, though, that any important churchman would urge that prayer be relied on exclusively. Even those with prayers on their lips are really counting on the doctors.

However much it may be that religion remains valuable as a system of ethics, its role as a controller of the Universe has been abandoned. And if the magic rites of the true religion are seen as useless, how much more useless must the magic rites of the false religions be?

But let's look at the situation from another standpoint. We speak of witchcraft, of the persecution of witches—yet witch refers to women. Although, in actual history, men have been denounced and victimized as practitioners of witchcraft—though wizards, warlocks, and sorcerers, have existed in plenty—it is not

---

* See "The Fateful Lightning" in *The Stars in Their Courses* (Doubleday, 1971).

as wizardry or sorcery that the practices are most commonly known. It is *witch*craft, and in the popular mind, the pre-eminent practitioner of witchcraft is the *witch* and she is clearly a *woman*. We have reached the point where the term "witch" is used almost exclusively, and where there is a strong temptation to define a practitioner of the male persuasion as a "he-witch."

In our present culture, we are most familiar with the witch as she is pictured in "Hansel and Gretel," in *Macbeth*, and in "Snow White" and as she is caricatured at Halloween. Always, she is an ugly old woman, with a curved nose and a curved chin approaching each other.

In the course of the witch mania of 1500 to 1750, though men and young women were tortured and killed, the percentage of old women persecuted was out of all proportion to the fraction of the total population they constituted.

Why? Was it male chauvinism, or was there some material difference between old women and old men, or between old women and young women, that made the old women particularly vulnerable?

Let's see.

To begin with, the human life-span was quite short prior to the present century. The average life expectancy was anywhere from twenty-five to thirty-five, depending on the time and the place. As a result, the percentage of old people was much lower than it is today, and that affected their role in society.

Because an old man was rarely found, he was valued when he was found. Because the chance of living a long time was greater if you were a member of the well-fed aristocracy, the percentage of old men who were of the upper classes was considerably out of proportion, and it was the easier to associate old men with rule.

In a preindustrial and, particularly, in a preliterate society, old men were peculiarly valuable for their memories. In the absence of the kind of records we have in written or electronic form, it was the brain of the aged man that was the repository of tradition and the final court of decision. Old men remembered how it was in the old days, and remembered the consequences of numerous decisions in the past. Because of this experience it was natural to have them rule the tribe, conduct the rituals, act as advisers. The very word "priest" is from the Greek word for "old" and the word "senator" is from the Latin word for "old," so that to this day we pay lip service, at least, to the notion of rule by the old.

Consider, too, that men have beards. In most early Western cultures, beards were a universal masculine adornment and were widely regarded as signifying manhood. Right into recent times a standard sneer at a young man was that of being a "beardless youth."

The beard, since it was a symbol of manhood was untouchable except by those closest in affection. To touch the beard was an insult, to pluck hairs out of the beard was a deadly offense. When Hamlet is trying to chide himself to action by imagining insults he says:

"Who calls me villain? breaks my pate across?
Plucks off my beard and blows it in my face?"

Again, when in the time of David of Israel, the King of Ammon ordered David's ambassadors to be forcibly shaved, that was grounds for war.

The verb "to beard," meaning to pluck at the beard, therefore (since it invites instant retaliation) also means "to set at defiance." Hence the phrase "to beard the lion in his den," meaning to have the courage to defy one on his home ground, where he is strongest.

The respect due the beard is magnified when the beard is white, since this marks not only the manhood of the wearer but the rare dignity and experience of age.

It seems to me, then, that all we know of earlier times leads us to suppose that old men, provided they maintained themselves with reasonable dignity, were reverenced.

And what about women?

To begin with, women differed from men, right down

153

to contemporary times, in having a much shorter life-span. They suffered the risks of famine, infection, and violence as man did, but on top of that they had to run the gantlet of childbirth. Women were baby-machines in those days and if one child did not kill her in its coming forth, the next one might.

In fact, it was not till the germ theory of disease was developed, and it became reasonably safe for women to have children, that women could live out a normal lifetime as a matter of course. It then turned out that they were more long-lived than men by five or ten per cent. She was the better biological specimen, thanks to an extra chromosome, once the threat of childbirth was removed.

But, in early times, when childbirth was the slaughterer of mothers and would-be mothers, old women were even rarer than old men.

Should old women, then, not be even more reverenced than old men? Perhaps not. In a male-dominated society, women are rarely, if ever, given a place in the ruling bodies of state and church. Their role was to have baby after baby and to stay indoors. Mere age, therefore, did not lend old women the valuable leadership qualities it lent old men.

Yet age and experience should lend old women some sort of specialized knowledge, shouldn't it?

Yes, of course. Since women spent their time in the society of infants and children and were required to amuse them with stories, they became the repository of folklore and of amusing beliefs that they lacked the experience and learning to see through. Self-righteous and self-superior man smiled at this and was always willing to maintain that women were more gullible, more superstitious, and more fearful than man so that it was only natural that they were all filled to their eyebrows with nonsense.

Therefore, while one could speak of old men's wisdom, one laughed at and scorned "old wives' tales." This is done even in the Bible, where the godly are warned to keep away from foolish superstition. "But refuse profane and old wives' fables, and exercise thyself rather unto godliness" (1 Timothy 4:7).

There's something else, too. Women, having to deal with children's illnesses, would pass on to each other a great many home remedies, most of which did no harm, at least, and helped psychologically. Sometimes the remedies were even legitimately valuable and an occasional woman would add discoveries of her own and pass those on.

An old woman was, therefore, very often the village doctor, and was the local expert in herbs, infusions, decoctions, and spells (like Grannie in "The Beverly Hillbillies").

The old woman, in her role as doctor, should have been respected—but she had to be feared as well. After all, spells and medicine can kill as well as cure and, with such powerful control over human life in her withered old hands, who could tell what an old woman might choose to do?

In this balance between respect and fear, fear won the day, in my opinion, for the very simple reason that a woman has no beard!

Remember that a good European beard can grow into a dense thicket that obscures virtually the entire face. This is important since in doing so, it obscures the ravages of age, except for the whitening of the beard itself, and that, after all, is a mark of reverence.

As a beardless woman ages, however, the wrinkles that form on her face are not hidden! An old woman looks very different from a young woman under these conditions, while an old man does not look so much different from a young man except for the color of the beard.

Combine the rareness with which one finds an old woman and the startling difference in her face as compared with a young woman or with a man of any age, and you have a powerful incitement to fear. Merely because she is so different and so rare, an old woman will seem ugly and repulsive—and frightening.

There is more, too. Consider the consequence of that

155

most common of all diseases, the one disease (apart from old age itself) that is almost universal, that strikes almost all people and is irreversible even today! If you haven't guessed, it's caries—tooth decay.

In earlier times, when sugar and all the sugar-riddled delicacies we eat today were rare or nonexistent, there was less tendency to decay. But there was some, and with no dental care except for the yanking out of teeth by main force when they ached too badly—life meant a steady drizzle of lost teeth.

To the average man or woman this didn't matter. With the average life-span something like thirty, some of the teeth lasted. To those lucky few who endured into old age, however, the price might be a total lack of teeth, and this is an age when false teeth were just about unknown.

The toothless old man would get away with it to a certain extent. His facial hair obscures his jaws and hides the effects of an edentate mouth.

Not so for the old woman. Her hairless face leaves her jaws in full view. Her teeth no longer keep the opposing gums apart, and when her mouth closes, the nose and chin come far closer together than they were wont to do, or than they ever do in young women generally. The approach of nose and chin became the mark of the old woman, therefore, and this is easily exaggerated into a downward hook to the nose and an upward hook to the chin in any caricature.

Do you begin to recognize the witch? Just a toothless old woman.

The toothless jaw carefully gumming soft food; the wrinkled face, wrinkling further as the jaws move together; the slurred speech with some sounds distorted by the absence of teeth; the ingratiating, toothless smile of a weak old woman—can all be terrifying when it is a rare phenomenon.

And she is helpless. Her husband and children are probably dead, because they have lived only the normal lifetimes of the day. If she has grandchildren, they may be indifferent to her. There is no feeling of any responsibility toward her on the part of society in general. — How can she live?

She might beg. More likely, she would play her role as village doctor, as midwife, as potion-dealer, as spellcaster. In order to make sure that her fees were as high as possible, and were duly paid, she would certainly exaggerate her own powers, and try to seem as powerful as possible. Who wouldn't for heaven's sake?

But how dangerous that must have been. She was bound to lose patients (even modern doctors do) and how would that be accounted for? The more powerful she was thought to be, the more unlikely it would seem that someone would die unless she had deliberately brought it about out of malevolence.

Then, too, people and domestic animals frequently sickened without apparent cause in that time of non-existent hygiene. The cause had to be demons or evil spells, and who could best control these things but the strange, old, ugly, wrinkled, mumbling herb-woman who claimed fearsome powers and who must surely have been teased by children or turned away by adults at some time—and who had then wreaked her revenge.

So whenever people began to hunt down witches, it was the old women who were bound to suffer out of all proportion.

And what has saved old women from this sad and brutal hounding? Kindly philosophy? The rule of religion and morality? Never!

Nothing vicious in mankind has ever been cured, in my opinion, by anything but the advance of science. I have said already that science removed the fear of witchcraft by offering an alternate method of controlling the Universe. And science, by wiping out the dangers of childbirth and extending the life-span, made old women much more common and therefore less strange and frightening.

Finally the advance of dentistry save natural teeth into old age or substituted false teeth of efficient design for those that were lost. With that, the mumbling, hook-nosed, hook-chinned caricature of feminine old age vanished. We have a new vision of old women now, one that differs very little from the vision of young women, after all.

So the wicked witch is dead—thanks to dentistry.

12

The Nightfall Effect

Once, many years ago, when I had just turned twenty-one, I wrote a story called "Nightfall," which, to my utter astonishment, turned out to be a classic.

It began with a quotation from Ralph Waldo Emerson's essays, one which John W. Campbell of *Astounding* had called to my attention. It went:

> If the stars should appear one night in a thousand years, how would men believe and adore, and preserve for many generations the remembrance of the city of God?*

Campbell wanted a story that would put that quotation in reverse and I was glad to oblige. I set my story in a world with six suns, which experienced nightfall under unusual circumstances only once in a long while and, when the stars appeared, everyone went mad.

I never thought that such a story had any predictive value at all. I could imagine my positronic robots coming to pass. I could even conceive of my all-human Galactic Empire being founded one day.—But intelligent beings going mad at the sight of stars?—And in my own lifetime?

It's happening. The thought of human expansion into space, the thought of man reaching for the stars, seems

---

* A few years ago, I admitted in print that I had never been able to find the source of the quotation. At once, dozens of letters arrived giving the precise reference in an Emersonian essay called "On Nature." So please, Gentle Readers, cease and desist from here on in. I now know.

to drive some people into an unreasoning frenzy. They begin to think of reasons against it and, in their madness, can come up only with mad reasons.

Let me be specific. A couple of years ago, Professor Gerard O'Neill of the physics department of Princeton University began to publicize his idea of establishing space colonies at points in the lunar orbit about the Earth, with the Moon itself serving as the source of the structural material.

I was skeptical at first because for years I have been banging away at the notion of colonizing the Moon itself. It took a while for me to see that I was suffering from what O'Neill called "planetary chauvinism"—the assumption that human societies must be built on, or just under, the surfaces of large worlds, just because that's where, by the accident of circumstance, the only technological society we know of happens to exist.

After my reading and thinking, however, he won me over. I became a convert to his views. (I may not be twenty-one any longer but I'm not so old as to have lost my flexibility.)

I began, therefore, to write articles boosting O'Neill's notion. In particular, I wrote one called "Colonizing the Heavens" which appeared in the June 28, 1975, issue of *Saturday Review*.

As a result of that article, I received a number of letters, some of them polite and some of them nasty, but all of them expressing the gravest doubts not only concerning the value of O'Neill's idea, but concerning the value of *any* notion of leaving Earth. Reading the letters saddened me, for there was not a sane attitude in the lot. Some of the objections were earnest and well-meant, but none were sane.

None of my correspondents, for instance, raised either of the two technical considerations that most tend to spoil the pretty picture of space colonization. Here they are:

First—The Universe is, by and large, a dangerous place for life because of the cosmic ray flux which, as far as we know, exists everywhere. Cosmic rays are highly penetrating, highly dangerous, and cannot easily be warded off or neutralized by anything manmade.

We get along on Earth because we have a planetary magnetic field which diverts some of the cosmic rays and miles of atmosphere which serves to absorb the worst of them. We would get along on the Moon, which has neither a magnetic field nor an atmosphere, because we would build a colony under meters of lunar crust and that would serve as adequate protection.

In a space colony, however, in which the walls are relatively thin and the atmosphere is on the inside, can the human inhabitants be protected from cosmic rays? O'Neill thinks that if the colony is large enough, it can be designed to absorb enough of the cosmic rays before they reach the people.

Second—A space colony has an insignificant gravitational field of its own and it is possible that, without a gravitational field, life would be uncomfortable and even, in the long run, impossible. In order to supply the equivalent of gravity, then, it is proposed that the colony be made to rotate rapidly enough to produce a centrifugal effect that will hold everyone to the inner surface of the curved wall of the colony with a force equal to Earth gravity.

A centrifugal effect in a comparatively small colony does not, however, truly duplicate the gravitational effect on a comparatively large world. The intensity of the centrifugal effect drops off rapidly as one rises up from the colony's inner wall, while the intensity of the gravitational effect falls off only very slowly as one rises up from the world's surface. We will be trading Earth's constant gravitational pull for the colony's highly variable one and that may mean trouble.

A related problem is that of the Coriolis effect, which is small but measureable on Earth, and which would be far more intense on the colony. In essence, it would mean that if you jumped upward, or threw something upward, you or it would come down in the same place.

The behavior of moving bodies would be quite different from that on Earth and that might mean trouble.

These were not the kinds of objections, however, that my correspondents raised. They raised others instead, quite foolish ones—foolish enough to make it seem that they are suffering from what I call "the Nightfall effect" (madness at the sight of the stars).

Some correspondents for instance, dismissed the whole thing as "science fiction" and were very indignant that the magazine and I should have pretended that we were dealing with actual science.

Alas, it was quite clear that these people knew that I was a science fiction writer and therefore, I suppose, felt that this was a good way in which to express their contempt for me and for the article at the same time.

In doing so, however, they revealed that they didn't know what science fiction is (except, perhaps, as "something written by Isaac Asimov").

Science fiction, at its most rigorous, deals with extrapolations from the present state of scientific theory and technological art. It assumes at least one advance that may or may not be possible, that may or may not ever take place, and then proceeds to tell a story.

The notion of space colonies, as advanced by O'Neill, does not do this, however. It makes use of on-the-shelf technology, of methods and techniques that are possible right now, and makes not one advance, however minor, beyond the present. In that respect, O'Neill's space colonies are not science fiction but are straightforward space technology.

Who says so? I, Isaac Asimov, the science fiction writer?

Of course not. I'm no expert in this. It's Gerald O'Neill who says so and his credentials are impeccable. He is not only a physics professor at Princeton, but he is highly regarded for his advances in nuclear research. It was he who conceived the idea of particle storage rings in which two accelerators send beams of particles crashing into each other head on, producing large increases in collision energy without the necessity of having to increase the size of the accelerators. (See his article on the subject in the November 1966 issue of *Scientific American*.)

It is, of course, possible for O'Neill to be mistaken. He is brilliant, but even brilliant people are human and can trip over their own feet. Yet his notion of space colonies has been published and discussed openly and at a number of conferences and it has withstood what one might call the heat of the kitchen. Just this morning (as I write) NASA has come out with the suggestion that a space colony (not quite in the form O'Neill proposed) be built in the course of the next half-century.

I would say, then that the notion is *not* science fiction.

Of course, my correspondents may have meant that space colonies were science fiction merely because they did not yet exist. That may be their definition of science fiction—a reference to anything that doesn't yet exist. In that case:

—A landing on the Moon was science fiction in 1968.

—The nuclear fission bomb was science fiction in 1944.

—The airplane was science fiction in 1902; and so on.

If that's science fiction, then I ask nothing better than to write science fiction articles for general magazines from now on.

A number of my correspondents felt indignant over the fact that I was raising false hopes over the expansion of the human habitat; that I was helping to convince people that mankind would soon spread out into space and that human beings generally need not feel, therefore, that there is any necessity to limit population. The population explosion would then continue and mankind would be doomed.

This bothered me, and I admit I blamed myself for not making it perfectly clear in my article that, in the short run, space colonization would not in the least affect the necessity of reducing the birthrate on Earth.

After all, if we build a space colony to hold ten thousand people over the next fifty years, what will this mean in comparison with the fact that at the present rate of increase there will be seven billion *additional* people on Earth by then. Subtract ten thousand from seven billion and you still have just about seven billion.

I am as aware of the danger of the population explosion as any man on Earth,† and yet the attitude of those who wrote me struck me as excessive to the point of irrationality. They apparently feared that any amelioration of the lot of mankind was bad because it would encourage a further population explosion. It seemed to me that they desperately wanted matters on Earth to grow rapidly worse as the only means of convincing people to take action. It was as though they were relying on catastrophe to prevent catastrophe.

But is that the only way out? The efforts, on the one hand, to control population and, on the other, to ameliorate the human condition, do not strike me as necessarily mutually exclusive. Let us consider an analogy—

Suppose a man is deeply in debt and must somehow gain the money to pay it off in the course of the year or face death. He has a certain income and if he cuts expenses to the bone, he may save enough to avoid death. But then suppose a friend comes along and points out to him that he might increase his income by taking certain actions.

Is this friend doing the debtor a disservice? Will the debtor, noting that he can increase his income, proceed to spend more than ever, thus actually increasing his debt and making his own death certain? Or will he welcome the increase and realize that he can cut his expenses and increase his income at the same time, and that by doing *both* he will make it more probable that he will live than if he concentrated on either strategy alone?

---

† See "The Power of Progression" in *The Stars in Their Courses* (Doubleday, 1971); "Stop!" in *The Left Hand of the Electron* (Doubleday, 1972); and *Earth: Our Crowded Spaceship* (John Day, 1974).

Well, which will he choose?

If he is stupid to the point of madness, he will allow the extra income to lure him into increased debt. If he is weak to the point of madness, he will refuse the added income for fear that it will lure him into folly. If, on the other hand, he is sane, he will see the value of the combined strategy.

The people who write to me are clearly convinced that human beings are, collectively, stupid and weak to the point of insanity. Well, maybe they are, but *if* they are then nothing will save our civilization and we might as well forget all about strategies. Therefore, we have nothing to lose if we assume, as an academic exercise, that human beings will, before it is too late, act rationally enough to save civilization and we ought to push for the most valuable strategy. That is to cut the birthrate as far and as fast as we can and, *at the same time,* to do what we can to maximize room, food, energy, and resources.

To be sure, my analogy is not quite accurate in that the debtor is pictured as knowing the desperateness of his situation whereas a large percentage of the Earth's population is either completely unaware that there is a population problem at all, or, being aware, chooses to deny its importance for a variety of reasons. But this is something to be countered by education in some fashion other than by the desperate invitation of catastrophe. If only catastrophe will educate, then humanity is mad and civilization will die—and I am assuming this is not so.

Some of my correspondents objected that the whole thing would be too expensive.

NASA suggests that the first space colony will cost a hundred billion dollars spread over the next fifty years and that's two billion dollars a year. But put that in perspective and you will see how irrational the expense objection is.

How much do Americans spend on alcohol and tobacco each year? Frankly, I don't know, but I'm willing to bet it's more than two billion dollars a year.

That is true, I think, if we just count the money spent on the actual physical drinks and smokes. What about the money value of the people killed and maimed by drunken drivers? What about the money value of the buildings burned down and the people killed by people who smoke in bed? What about the money value of the forests burned down by smoking campers? What about lung cancer and heart attacks due to smoking?

Add it all up. If we can spend astronomical amounts on habits known to be deadly, both to those who indulge and to many who do not, isn't it insane to object to spending less on a program that may do mankind infinite good?

You may want to argue that individual people enjoy smoking and drinking and won't give it up for colonies in the sky, and that it is cheap for me to condemn the habits, when I don't possess them and therefore don't know what I'm missing.

In that case, consider warfare. How much money has been spent (just money; never mind lives and such abstractions as human happiness) on wars thus far in the twentieth century? How much money has been spent and is being spent on preparations for war?

Suppose we develop a world without war. The money saved will pay for the program of space colonization many times over, and we won't even speak of the lives saved and the misery abolished.

Then, too, remember that the price of war (and of alcohol and tobacco, too, for that matter) has been going up steadily each decade through all modern history. The price of space colonization is, on the other hand, likely to go down. More and more the Moon will be used as the quarry, space as the work arena, the colonists as the workers. More and more the colonists will return more to Earth (in the form of beamed solar energy, for instance) than they take.

But is all this hogwash? Is it impossible that humanity will give up war and disband its armies and, therefore, that it will ever have the money to spare for space? In that case, it is also impossible that we will move very far into the twenty-first century with our technological civilization intact. No strategy will save us and we may as well amuse ourselves by aiming high.

Some felt that space colonies would be a fiasco since no one would want to live in an engineered environment.

Does one laugh or weep at that piece of madness, since no one can possibly maintain that view without carefully ignoring every scrap of human history?

The first cities were built about ten thousand years ago and every succeeding decade since has seen the Earth, on the average, more urbanized. There have been declines in urbanization during dark ages, but dark ages have always been localized (at least till now) and the overall movement has remained upward As an example of something recent, some fifteen per cent of the American population lived in urban areas in 1900, but in 1970 the figure was eighty-seven per cent.

Clearly, urban areas are more engineered (more artificial, more removed from the original natural state, more dependent on a complex technology) than rural areas, and the level of engineering in cities has been steadily rising. Far from people rejecting engineering, then, all history of the last ten millennia shows humanity to be rabidly hungering for ever more engineering. Today the movement toward the cities is greater than ever.

In the United States, some of you may say, there is a flight *away* from the cities. Yes, indeed, there is—a flight from the *central* cities, where the engineering is obsolete and breaking down. The flight is to the city's outskirts, the suburbs, where the engineering is efficient. The flight is not away from engineering, but toward better engineering.

Is there a movement back to the soil? There always is, in every generation—a tiny (but noisy) ripple in a tidal wave.

How is it possible then to say that no one would want to live in an engineered environment?

Some felt that it was the location that mattered; that no one would want ot leave Earth and go live in a space colony.

Mind you, those who write this in letters to me are Americans and presumably have some bare-bones knowledge, at least, of American history. And in that case, how can this suggestion be anything but the product of irrationality?

Imagine that someone makes the following proposition to you—You are to get onto a small sailing vessel, with bad food and water and with no privacy. You are to spend six weeks pitching and tossing on an ocean where the first real storm will sink you. If you survive and make a landing, you will find yourself in a trackless wilderness amid a hostile native population. Would you go?

Many people did. That was how the American colonies were settled in the 1600s. Some eleven thousand people came to the new colony of Virginia between 1607 (when it was founded) and 1617. Of these, ten thousand died, leaving only one thousand survivers in 1617. yet people continued to come.

Or imagine this proposition— You are to take a crowded steamship that will take a week or more to make the voyage. You will travel in steerage and end up in the slums of a crowded city working in a sweatshop. Would you go?

Millions did throughout the 1800s and very early 1900s and filled the United States with every ethnic group of Europe.

Or imagine this proposition— That you get into a wagon, or that you possibly go on foot, and cross fifteen hundred miles of untamed land, some of it desert, and all of it subject to incursions by hostile natives. For your pains you will reach territory that may or may not contain gold, which you may or may not find after infinite pains.

A number did exactly that in 1849 and afterward. And some joined a similar mad rush a half-century later to seek for gold mines in polar Alaska and Canada.

Not want to go? All history shows us that when an old life has become unbearable, people will dare any danger and go to any lengths in order to find a new life and make a new start.

It seems almost inevitable that over the next fifty years, while the people of Earth try desperately to stop population growth, life will become steadily less bearable.

Not want to go then? With ten thousand posts available in the first space colony, I suspect that ten million will volunteer.

Some looked forward to a future where space colonies were plentiful and they feared that the educated, sophisticated, and intelligent would leave Earth—the beautiful people Earth couldn't afford to lose. They felt that the colonization of space would leave Earth a human wreckage heap.

These, too, forget American history with a thoroughness that seems to require irrationality.

Why should the educated, the advanced, the well-off want to leave Earth for the space colonies? They would be comfortable here. Was it the educated, the advanced, the well-off who flocked into the cockleshell sailing vessels and the stinking steerage of steamboats to get to the New World? Was it the educated, the advanced, the well-off who left the cities of the East for the gold mines of California?

No, sir. Those who came were not the British peerage, but the starving Irish peasants; not the Tsar's entourage, but bedraggled Jews from the ghettos. Yes, some scholars came, but the vast majority were those who were so badly off at home that not all the dangers and difficulties of the trip, not all the hardships in the new land, would keep them away.

There is Emma Lazarus's poem inscribed on the Statue of Liberty to make the point. Part of it goes—

... Give me your tired, your poor,
Your huddled masses yearning to breathe free,
The wretched refuse of your teeming shore,
Send these, the homeless, tempest-tossed, to me:
I lift my lamp beside the golden door.

Perhaps I remember this better than those who wrote me letters, because my parents and I were part of the wretched refuse. We landed on Ellis Island in 1923, one year before the golden door was shut.

We're not going to lose the best people to the colonies. It's the poor ones who will clamor to go. We will have to bribe and browbeat the fat cats, if we want any of them to go.

Some feared that the colonies would end up racist—that people of the underdeveloped nations might want to go but couldn't, lacking all ability at space engineering and having no experience with space.

How can anyone suggest this even in a fit of madness?

My parents (to get back to them) had never seen the ocean till they made their trip to New York. They had never seen an ocean liner till they boarded one. And after they boarded it, do you think they had the slightest notion of how it worked, or even of how it floated? That didn't stop them from going to the United States.

For heaven's sake, is it difficult to understand that to go from Earth to a space colony, one does *not* need to be a space engineer, one does *not* have to pilot a spaceship, one does *not* need any experience with space?

What one needs to get there (hold your breath, now) is a ticket.‡

So there you have the Nightfall effect. Some people faced with the stars go mad. What other conclusion

‡ People might not be able to afford a ticket, but that's a different thing, and is something to be solved by an economically sane society, which I hope will be developed by the twenty-first century.

can you draw from such arguments as they have presented?

It may be that by the time we enter the twenty-first century our technological civilization will be irretrievably falling apart. If so, we will not go out into space; we may never go out into space.

But let's suppose we do survive into the twenty-first century. In that case, in a low-birthrate world without war, space *will* be explored and colonized and the stage *will* be set for a new and greater expansion of mankind to a new and far higher level of civilization.

I won't live to see it in actuality, but that doesn't matter at all, for I see it in my mind's eye, and—providing only that our civilization survives—*I know it will be so!*

13

The Rocketing Dutchmen

Quite frequently I get books, magazines, and miscellaneous printed material in the mails, stuff I haven't requested and didn't expect. My first impulse in such cases is to look at the index, if there is one, or riffle the pages if there is no index, to see if my name is mentioned. Such mention is often (not always, however) the reason the material is sent to me.

When the object in question is a subject on which I have expressed myself in some sardonic way, I am particularly suspicious. For instance, a rather considerable time ago I got something called "UFO Symposium—1973," and in it was an article by Stanton T. Friedman, a gentleman with whom I am not acquainted.

The article contained a section called "Science Fiction Vs. Ufology," which begins, "Many people are surprised when I point out that two of the most noted science fiction and science writers, Isaac Asimov and Arthur Clarke, are both quite vehement in their anti-UFO sentiments."

That Friedman meets people who are "surprised" at this indicates, I suppose, the level of the circles he moves in. After all, why should the fact that Arthur and I are s.f. writers lead people to suppose that we have forfeited our intelligence and must surely believe any mystic cult that seems to have some elements in common with science fiction?

Friedman goes on to quote me and to add his own asides, designed, I presume, to smash me into silence. Thus he quotes me as saying, "The energy requirements for interstellar travel are so great that it is inconceiv-

172

able to me that any creatures piloting their ships across the vast depths of space would do so only in order to play games with us over a period of decades. If they want to make contact, they would *make* contact; if not, they would save their energy."

To this, Friedman says, in parentheses, "(What egos we earthlings have! Are we worth contacting?)."

Friedman has obviously quoted me without reading the quote. I said, "*If* they want to make contact—" I am perfectly ready to admit that we may not be worth contacting, but in that case "they would save their energy"—and go away.

Imagine the ego of the Friedmans who believe that perhaps we are not worth contacting but that we are nevertheless so fascinating that somehow the flying saucers keep nosing about our planets by the thousands over a period of decades, like rocketing Dutchmen doomed to circle Earth forever without landing, and condemned, further, to keep displaying themselves to us like male pigeons in heat.

Friedman then quotes a statement of mine that concludes: "I will continue to assume that every reported sighting is either a hoax, a mistake, or something that can be explained in a fashion that does not involve spaceships from the distant stars."

And Friedman, assuming a jocose familiarity, says, "(How about the nearby ones, Isaac?)."

Alas, Mr. Friedman,* even the nearby stars are distant.

Friedman goes on to urge me to write a nonfiction book about flying saucers, saying that "cases like the Betty and Barney Hill case are far more exciting and interesting than any of Asimov's stories." Well, perhaps, Mr. Friedman, but they are also much more fictional.

But if not a book, I shall write an article on the matter. Goodness knows, I have stated my views on flying saucers a number of times, but I have never done so in

---

* I myself prefer not to assume a familiarity that does not exist.

an article of this series. Let me do so now in question-and-answer fashion:

1) *Why do you insist on calling them "flying saucers"? Isn't that unfair ridicule? Why not call them UFOs, a more sober term?*

UFO stands for "unidentified flying object." If I discuss the matter with someone who agrees that these manifestations, whatever they may be, are, in fact, unidentified, and does not insist on identifying them, then I will gladly discuss UFOs as soberly as possible. To anyone, however, who insists on identifying them as spaceships piloted by extraterrestrials, the objects are *not* unidentified and are therefore *not* UFOs. In that case I call them flying saucers, which is the term the flying saucer enthusiasts themselves used before they decided to try for respectability.

2) *Do you deny that there are other intelligent life forms in the Universe?*

I certainly do not deny that. As long ago as September 1963, I wrote an *F & SF* article entitled "Who's Out There?" in which I followed the arguments of Carl Sagan to the effect that there could be numerous civilizations in the Universe.

Then, in collaboration with Stephen H. Dole, I wrote a book, *Planets for Man* (Random House, 1964), which took up the matter in greater detail and from a slightly different viewpoint and which advanced the suggestion that there were numerous life-bearing planets in the Universe.

Let me repeat that argument very briefly—

No one really knows how many galaxies there are in the Universe; certainly many billions. A hundred billion is the figure I usually use. Even if we confined ourselves to one galaxy only, our own Milky Way Galaxy, we still have a system that contains 135 billion stars.

Current theories of stellar formation suggest the invariable formation of planetary systems when a star is born, so we can say our Galaxy contains 135 billion planetary systems, each containing perhaps a dozen planets and half a dozen large satellites.

Of these better than a trillion cold bodies, some are too far from their star to be Earthlike, some too near. Some might have rotations that are too slow, or orbits that are too eccentric, to allow a comfortable weather pattern. Some might circle stars which are too cool to supply the necessary energy for life, or too hot and therefore too short-lived to give life the necessary time to evolve. Some might circle stars that are parts of multiple systems, or stars that pulse, or stars that, in other ways, make the environment too uncomfortable.

Even taking all this into account, Dole, making fair estimates in the light of the astronomy of the early 1960s, concluded that there might be as many as 640 million Earthlike planets—planets with roughly Earth mass, Earth temperature, and Earth chemistry; and with an Earthlike orbit and an Earthlike Sun—in our Galaxy.

This is not too generous an estimate since it means that only one planetary body out of four thousand is suitable, and that only one star out of about 210 has an Earthlike planet.

Yet perhaps it *is* too generous if we take into account astronomical developments of the last decade. Since about ninety per cent of the stars of the Galaxy are in the galactic nucleus, some ninety per cent of the Earthlike planets ought to be there, too, if we assume even distribution.

The nuclei of galaxies may, however, be the scenes of violent activity—quasars, explosions, black holes, etc.—and it may only be in the spiral arms of a galaxy (where we are) that conditions are quiet enough for planets to be truly Earthlike. In which case, we might perhaps estimate only 64 million Earthlike planets in our Galaxy.

However, the more Earthlike planets there are, the better the case for flying saucers, so let's be generous and keep the larger figure of 640 million.

By current theories of the origin of life, any planet that has an Earthlike environment will inevitably develop life. What we are saying then, is that there may be

175

640 million life-bearing planets in our Galaxy—and life more or less as we know it, too.

Now comes the point at which speculation becomes thinner. On how many of these life-bearing planets does an intelligent species develop, and on how many does this intelligent species develop a civilization?

The only thing we can use as a starting point is Earth itself, the one life-bearing planet we actually know. On Earth, life has existed for some three billion years and civilization has existed for, at most, some ten thousand years. This means that noncivilized Earth outstretches civilized Earth by 300,000 to 1.

If we assume Earth to be average and this to be a general rule, and that life started at different times in different places, we can assume that civilization exists on one out of every 300,000 life-bearing planets. In that case we have about 2,150 civilizations in our Galaxy.

As for an *industrial* civilization, we have had one for two hundred years out of our ten thousand years of civilization. In other words, our nonindustrial civilization outweighs our industrial technology by 50 to 1.

If we suppose that one out of fifty civilizations in our Galaxy has reached the industrial stage, then there are some forty-three industrial civilizations in our Galaxy.

If we further suppose that our own industrial technology is about average, as such things go, then half of these industrial civilizations—say twenty-one—are more advanced than ours and are capable, perhaps, of space travel.

That's just in our Galaxy. If this sort of reasoning holds for all galaxies, then there are, perhaps, as many as two trillion advanced civilizations in the Universe. But then, I suppose even the most convinced flying saucer enthusiast would agree to eliminate other galaxies as the source of our visitations and be willing to confine himself to our own single Galaxy. That would still leave twenty-one possible civilizations wandering around the footless halls of space, and surely these are enough to account for flying saucers, if flying saucers are space-ships.

3) *Well, then, why are you so skeptical of the pos-sibility that spaceships guided by extraterrestrial intel-ligences are visiting Earth?*

For one thing, the distances disturb me. Imagine all the 640 million life-bearing planets distributed random-ly through the Galaxy. They would then be, on the average, about forty-five light-years apart. The twenty-one planets with advanced industrial civilizations on them would be, on the average, 13,500 light-years apart.

With the nearest home planet of flying saucers 13,500 light-years away, the chance of visiting us would seem small.

Since the speed of light is the limiting speed at which a spaceship can move toward us, one coming to us from even the nearest advanced civilization would take 13,500 years (by stay-at-home time on their native world) to reach us and, very likely, ten times that long. It seems doubtful to me that, under those circumstances, ship after ship would buzz around us, for year after year, like bees around clover. We can't be either that inter-esting or important.

4) *But suppose we just happen to be in luck as far as the distance of the nearest advanced civilization is con-cerned? And why are you so certain that the speed of light is the ultimate limit?*

I don't insist on being categorical about such things. Assuming random distribution, some advanced civiliza-tions may clump together, some be fearfully isolated. It may just happen that Earth is only one hundred light-years from a very advanced civilization. This would be tremendously unlikely, but there is no evidence one way or the other and it *may* be so.

Then, too, even if the original centers of the civiliza-tions are far, far apart, and if none is particularly close to us, each may nevertheless be the nucleus of a grow-ing Galactic Empire, and there may be outposts of some

Empire on some of the nearer stars. There is no evidence for this either, but it *may* be so.

Then, too, perhaps advanced civilization may learn to circumvent the speed-of-light limit without violating relativity. Perhaps they can learn to make use of hyperspace or of a tachyonic drive or something that we, in the feeble state of our own technology, can't put words to, or have concepts for. This doesn't seem very likely, actually, but it *may* be so.

Perhaps, then, distance is unimportant to the advanced civilizations. Perhaps they can cover a hundred light-years or even 13,500 light-years with no greater difficulty than we can fly across the Atlantic Ocean.

5) *But if all that is so, what are your objections to the concept of flying saucers? Why might not ships be exploring Earth freely and frequently?*

If we ignore the question of distance, there remains that of motive. If these rocketing Dutchmen are buzzing around Earth deliberately and for some rational reason, it must be because Earth interests them. But what on Earth can possibly interest them?

It is natural (if perhaps egotistic) to assume that to any outworlder the most interesting thing about Earth is man and his civilization. But if the flying saucers are investigating us, why don't they come down and greet us? They should be intelligent enough to work out who our spokesmen are and where our centers of population are and how to go about making contact with our governments.

Nor is it conceivable they can be afraid of us. If their technology is such that they can cover multi-light-year distances without trouble, then they can easily protect themselves against any puny weapons we can turn against them. Would an American warship be afraid to land an exploration party on an island occupied by chimpanzees?

If there is something about our atmosphere or our surface that they find deadly or just unpleasant, they are surely intelligent enough to communicate with us by some sort of long-distance transmission—radio, if nothing else. If not words and language then some signal obviously born of rationality.

On the other hand, if they are interested in us, but do *not* wish to make contact with us—if they do not wish to interfere in any way with a developing civilization—they are certainly intelligent enough and advanced enough to be able to study us in whatever detail they need without ever letting us be aware of them. By letting us be aware of them, they *are* interfering with us.

And if it is something else than man that interests them, then what?—No, they should either come down and say hello, or they should go away. If they do neither, they are not intelligently guided spaceships.

6) *But how can you be sure you understand their motives? Perhaps they don't care to communicate with us, but, on the other hand, don't care if we see them.*

Ah, but if you keep on piling up the conditions you need to improve your case, you come very rapidly to the point of the totally unconvincing.

To get rid of the object of distance, you must *assume* at least one civilization improbably near to us, and you must *assume* the achievement of faster-than-light travel. To get rid of the puzzle of their behavior, you have to *assume* that they find Earth interesting enough to pester repeatedly, but ourselves so uninteresting they won't talk to us, while, on the other hand, they don't care if we see them.

The more assumptions of this sort you must make, the weaker your case.

Actually, none of these assumptions have any support, whatever. The only function they serve is to explain flying saucers. One can then use the flying saucers themselves as an argument to say that the assumptions must be correct. This is arguing in a circle, one of the chief delights of the intellectually feeble.

7) *Now wait, there is certainly direct evidence for flying saucers as spaceships. There have been numerous*

reports from people who have seen spaceships and their extraterrestrial crew members. Some claim even to have been aboard the ships. Have you investigated these reports? If not, do you dismiss them all out of hand as worthless? What justification do you have to do that?

No, I have not investigated any of these reports. Not one.

My justification in dismissing them out of hand is that eyewitness evidence by a small number of people uncorroborated by any other sort of evidence is worthless. There is not a single mystical belief that is not supported by numerous cases of eyewitness evidence.

There is eyewitness evidence (as reported by enthusiasts) for angels, ghosts, spirits, levitation, werewolves, precognition, fairies, sea serpents, telepathy, abominable snowmen, and so on, and so on, and so on.

I won't throw myself into the morass of believing all these things on eyewitness evidence alone; and if I don't, I won't believe flying saucer spaceships on eyewitness evidence alone either. I want something less prone to distortion, and less subject to deliberate hoaxing, than eyewitness evidence is.

I want something material and lasting, something that can be studied by many. I want an alloy not of Earth manufacture. I want a device that does something by no principle we understand. Best of all I want a ship and its crew in plain view, revealing itself to human beings competent to observe and study them over a reasonable period of time. These reported revelations to farmers in swamps and to automobile drivers on empty highways simply don't impress me. Nor am I impressed by descriptions of the ships and their interiors that are what I would expect from scientific illiterates who had seen some equally illiterate "science fiction" movies.

8) *But how else can you account for all the reports of flying saucers if you're going to rule out spaceships?*

There is the well-known Holmesian dictum that "Whenever you have eliminated all that is impossible, whatever remains, however improbable, must be true."

That is a great fraud, for it presupposes that after the elimination of the impossible, you are left with only *one* remaining factor. But how can you know that?

This misconception has arisen out of mathematics. In mathematics, we can so organize our definitions and axioms that we may be presented with a small number of factors and no more, with every one of that small number known. In that case if we eliminate all but one, the remaining one must be true (provided we show that it is not possible for none to be true).

This does not apply to the experiment or observational sciences, where the total number of factors may be indefinite and where not all may be known.

If flying saucers are spaceships, this must be proven by direct evidence. It can never be proven by wailing, "But what else can it be?"

9) *What do you think yourself that flying saucers are?*

My own feeling is that almost every sighting is either a mistake or a hoax. Many are so confused and incomplete there is no room to decide what they can possibly be.

I am told that there are some reports (a very small minority of the total) that seem to be neither mistakes nor hoaxes; that have been checked by reliable and keen observers; and that cannot be explained in an ordinary way.

10) *All right, stick to those puzzlers. What are they if they are not spaceships?*

I don't know. I don't have to know. The Universe is full of mysteries to which I don't have the answer. Challenging me and having me fail proves nothing.

Look, you may, perhaps, not know the name of the fifteenth president of the United States. If I say his name was Jerome Jameson, the fact that you don't know anything to the contrary doesn't prove my case.

But let us consider Joseph Allen Hynek, a respected American astronomer whom I know personally and who,

I can testify, is an honest and an intelligent man of thorough scientific attainments.

Hynek is not ready to dismiss flying saucer reports out of hand as most astronomers do (and as I myself generally do). Rather he wants them examined carefully, and he is doing so himself. It is not an easy thing to do. These reports are so riddled with hoaxes, and the flying saucer enthusiasts have so many cranks, freaks, and nuts among them, that Hynek is constantly running the risk of innocently damaging his reputation by being confused with them. His interest in these strange reports, however, and his belief in their importance is enough to make him willing to accept the risk and I honor him for it.

Hynek does not believe that the reports deal with extraterrestrial spaceships. He does not have an explanation ready to hand for the reports. With him the subject of discussion is UFOs, Unidentified Flying Objects.

What Hynek says is that there is something there; something that cannot be explained within the conventional structure of science; and something, therefore, that should not be ridiculed and dismissed, but should be carefully and thoroughly studied.

He thinks that the manifestations that cannot be explained represent something so new to science that when solved it will lead to an enormous advance; a quantum leap, he says.

It's happened before. The puzzle of the negative result in the Michelson-Morley experiment led to the quantum jump of relativity. The paradoxes of blackbody radiation led to the quantum jump of the quantum theory itself. Therefore it may be that the UFO puzzle will lead to—what?

It's a fascinating thought. Almost, Hynek persuades me.

11) *Does Hynek have any theories about this at all? Where does he think science may be heading?*

As far as I know, he's drawn a complete blank so far. He has spent a great deal of time in checking reports, in classifying them, and in seeking factors that various types of reports have in common, but when he's all through, he has a puzzle on his hands for which he has no answer.

12) *What makes it so difficult to find an answer to this problem?*

The scientific attack on the puzzles of the Universe works well when the system being studied is steadily available for either observation or experimentation or both. The planet Mars is usually available for telescopic study. A turtle heart is usually available for experimentation.

The scientific attack works well, too, when you can set up simple experiments, whose general tenor you understand. If you don't understand the underlying manner in which balls fall, you may nevertheless always set up any number of balls that fall under controlled conditions and study the results.

On the other hand, think of those relatively few UFO reports that are genuine puzzles and that are not either mistakes or hoaxes. Those UFO phenomena appear unheralded, unexpected, and with the utmost irregularity in space and time. There is no way of laying a trap for them, short of setting up a world-wide monitoring system that would be fearfully expensive.

When a UFO phenomenon appears, it may not be witnessed at all; or it may be witnessed only in part by an individual or a few individuals who are caught by surprise and who may have no chance to make careful observations and no equipment to do it other than by eye. We end with an anecdotal half-memory of something half-seen.

Furthermore, once a report of this sort is made, it is widely reported in the newspapers, and that means it is at once buried in innumerable similar reports retailed by honest unsophisticates, by eager publicity hounds, and by sick-minded hoaxers.

Under such conditions, it is not at all surprising that

Hynek can't find a solution easily. I would not be surprised if neither Hynek nor anyone else could find a solution—ever!

And one last point. I am afraid that Hynek's feeling that the solution to the problem would lead science through a quantum jump is just his *belief*. I don't blame him for his enthusiasm; I am myself riddled with various enthusiasms; but enthusiasm must be recognized for what it is and not mistaken for evidence.

I, myself, suspect (and it is just a suspicion) that if each puzzling UFO report were subject to thoroughgoing investigation, then the more that would be found out about it, the less puzzling it would seem. I believe that if all UFO reports were completely understood, all would turn out to be something that is part of our present structure of science or is, at most, an interesting but not very important amendment or extension of that structure. The solution of the UFO problem would add, I suspect, very little, or perhaps nothing at all, to science.

If I am wrong, and Hynek is right, I would be happy, for I like Hynek and I would be pleased to see science advance—but I can't make myself accept something just because it would please me to accept it. I have to accept only what seems to make sense to me.

14

Best Foot Backward

In my more self-pitying moods, I feel more and more as though I alone am defending the bastions of science against the onslaughts of the new barbarians. Therefore, although I may be repeating bits and pieces of statements I have made in previous articles, I would like to devote this one, in its entirely, to such a defense, which, I warn you, will be an entirely uncompromising one.

Item 1—You would think that in a publication like *New Scientist,* an excellent British weekly devoted to articles on scientific advance, there would be no space given to simpering antiscientific idiocy.—Not so!

In the May 16, 1974, issue, one of the magazine's feature writers, having delivered himself of a fairly incoherent defense of Velikovsky, went on to say: "Science in its 200-year flight has produced some neat tricks like canned food and long-playing records, but, truthfully, how much else of real value to man's three-score years and ten?"

I promptly wrote a letter in which I said, in part: ". . . one thing you might consider to be of real value *is* man's threescore years and ten. . . . Through most of history it has been more like onescore years and ten. May we expect a bit of gratitude from you for those extra forty years of life you have the chance of enjoying?"

The letter was published and, in no time at all, in the July 11, 1974, issue there came a blast from a gentleman from Herefordshire whom I shall call B. He felt, apparently, that longer life had its disadvantages, since it helped bring on the population explosion, for instance. He said also: ". . . those benighted times Mr. Asimov mentions that had a life expectancy a good deal less than threescore and ten did still manage to produce Chartres, Tintern, Raphael, and Shakespeare. What are the modern equivalents? —Centre Point, Orly, Andy Warhol, and SF?"

Noting the dig at science fiction and guessing from whom he meant to draw blood, I felt justified in removing the velvet gloves. In my answer I said, in part: "B. goes on to point out that short-lived men in centuries past produced great works of art, literature, and architecture. Is B. advancing this as an odd coincidence, or does he maintain the cultural advance of the past came about *because* men were short-lived?

"If, indeed, B. resents the extended lifetime science has made possible, and finds it destructive to humanity, what does he suggest? It would not be difficult, after all, to abandon the advances of science, to allow sewage to creep into our water supply, to eschew antiseptic surgery, to give up antibiotics, and then to watch the death rate rise to a level that will quickly produce (by B.'s novel line of argument) another Shakespeare.

"Should B. indeed welcome this, would he recommend that the benefits of a heightened death rate be applied only to the benighted heathen of other climes, the lesser breeds of darker color whose accelerating rate of passage might then make the globe more comfortable for the men of Herefordshire? Or does his rigid sense of fairness cause him to recommend that all nations, his own included, participate in this noble endeavor? Does he, indeed, intend to set the example himself by manfully and nobly refusing to have his own life extended by science?

"Has it in fact occurred to B. that one answer to the population explosion brought on by the advance of science and medicine is to lower the birthrate? Or does he, perchance, find the lowering of the birthrate repugnant to his sense of morals, and does he much prefer the glamour of plague and famine as a cure for over-population?"

That letter too was printed and there came no answer.

Item 2—I get private communications, sometimes, that express an individual's dissatisfaction with the modern world of science and technology, and call for a quick retreat, best foot backward, into a preindustrial world of nobility and happiness.

For instance, a letter arrived recently from a professor of something or other who had gotten himself a farm and was growing his own food. He told me jubilantly all about how great it was and how healthy and happy he felt now that he was freed of all that horrible machinery. He did use an automobile, he admitted, and he apologized for it.

He didn't apologize for the fact that he used a typewriter, however, and that the letter got to me by way of our modern system of transportation. He didn't apologize for the use of electric lights or the use of the telephone, so I assume he read by the light of a wood fire and sent messages by semaphore.

I simply wrote back a polite card wishing him all the joy of the medieval peasants, and that elicited a pretty angry reply that enclosed an unfavorable review of my book *Asimov's Annotated "Paradise Lost."* (Ah, yes, I remember now, he was a specialist on Milton and I think he objected to my invasion of the sacred precincts.)

Item 3—Once, during the question-and-answer session that followed one of my talks, a young man asked me if I honestly believed science had done anything to increase man's *happiness.*

"Do you think you would be just as happy if you had lived in the days of ancient Greece?" I asked.

"Yes," he replied firmly.

"How would you have enjoyed being a slave in the Athenian silver mines?" I asked with a smile, and he sat down to think a bit about that.

187

Or consider the person who said to me once, "How pleasant it would be if only we lived a hundred years ago when it was easy to get servants."

"It would be horrible," I said at once.

"Why?" came the astonished answer.

And I said, quite matter-of-factly, "We'd be the servants."

Sometimes I wonder if the people who denounce the modern world of science and technology are precisely those who have always been comfortable and well off and who take it for granted that in the absence of machinery there would be plenty of people (*other* people) to substitute.

It may be that it is those who have never worked who are perfectly ready to substitute human muscles (not their own) for machinery. They dream of building the Chartres cathedral—as an architect and not as a peasant conscripted to drag stones. They fantasy life in, ancient Greece—as Pericles and not as a slave. They long for Merrie Olde Englande and its nut-brown ale—as a Norman baron and not as a Saxon serf.

In fact, I wonder how much upper-class resistance to modern technology arises out of a petulant dissatisfaction over the fact that so many of the scum of the Earth (like me, for instance) now drive automobiles, have automatic washers, and watch television—thus reducing the difference between said scum and the various cultured aristocrats who moan that science has not brought anyone happiness. It has diminished the grounds of their self-esteem, yes.

Some years ago there was a magazine named *Intellectual Digest,* which was run by very nice people but, alas, didn't survive for more than a couple of years. They had run some articles denouncing science and felt that perhaps they ought to publish an article supporting science—and they asked me to write one.

I did and they bought it and paid for it—and then never published it. I suspect (but do not know) that they felt it would offend their clientele, who may have been, for the most part, members of that branch of soft-core intellectualism that considers it clever to know nothing about science.

That audience was, perhaps, impressed by an article by Robert Graves which was reprinted in the April 1972 issue of *Intellectual Digest* and which seemed to argue for the social control of science.*

Graves is a classicist, brought up in the British upper-class tradition in the years prior to World War I. He knows a great deal more about pre-Christian Hellenism, I am sure, than about post-industrial science, which makes him a dubious authority on the matter of scientific discovery, but this is what he says:

"In ancient times, the use of scientific discovery was closely guarded for social reasons—if not by the scientists themselves, then by their rulers. Thus the steam engine invented in Ptolemaic Egypt for pumping water to the top of the famous lighthouse on the Island of Pharos was soon abandoned, apparently because it encouraged laziness in slaves who had previously carried waterskins up the lighthouse stairs."

This, of course, is purest horseradish. The "steam engine" invented in Ptolemaic Egypt was a pretty little toy that couldn't have pumped water one foot, let alone to the top of the Pharos.

Yet never mind that. Graves' cautionary tale is true in essence even if it be false in detail. The Hellenistic Age (323-30 B.C.) did indeed see the bare beginnings of a kind of industrial age, and that this advance bumped to a quick halt may have been, in part at least, because slave labor was so available that there was no great demand for machines.

In fact, it is even possible to present a humanitarian argument against industrialization to the effect that if machines replaced slaves, what would one do with all the surplus slaves? Let them starve? Kill them? (Who says aristocrats aren't humane?)

---

* So am I, provided the control is exerted by those who know something about science.

189

Graves, then, and others like himself seem to be pointing to the social control of science in ancient times as being directed toward the preservation of slavery.

Is this, indeed, what we want? Are all the antiscience idealists to march bravely into battle under the banner of "Up with Slavery"? Or, since most antiscience idealists think of themselves as artists, gentlemen farmers, philosophers, or whatever, and *never* as slaves, ought the banner read "Up with Slavery for Other People"?

Of course, some deep thinker may point out in rebuttal that the kind of factory life made possible by modern technology is not better than the lot of the ancient slave. Such arguments were used prior to the American Civil War to denounce the hypocrisy of free-state abolitionists, for instance.

This is not an altogether foolish argument and yet I doubt that any factory hand in Massachusetts would have voluntarily agreed to be a Black farm hand in Mississippi under the impression that the two professions were equivalent. —Or that a Black farm hand in Mississippi would have refused to become a factory hand in Massachusetts because he felt that was no improvement over slavery.

John Campbell, the late editor of *Analog Science Fiction,* used to go further. He believed (or pretended to believe) that slavery had its good points and that everyone was a slave anyway. He used to say, "You're a slave to your typewriter, aren't you, Isaac?"

"Yes, I am, John," I would reply, "if you want to use the term as a metaphor in my case and as a reality in the case of a Black man in the cotton fields of 1850."

He said, "You work just as long hours as the slaves did, and you don't take vacations."

I said, "But there's no foreman with a whip standing behind me to make *sure* I don't take vacations."

I never convinced him, but I sure convinced myself.

There are people who argue that science is amoral, that it makes no value judgments, that it is not only oblivious to the deepest needs of mankind but entirely irrelevant to them.

Consider the views of Arnold Toynbee, who, like Graves, is an upper-class Englishman who spent his formative years before World War I. In an article in the December 1971 issue of *Intellectual Digest,* he said: "In my belief, science and technology cannot satisfy the spiritual needs for which religion of all kinds does try to provide."

Please note that Toynbee is honest enough to say "try."

Well, then, which would you prefer, an institution that does not address itself to spiritual problems but solves them anyway, or an institution that talks about spiritual problems constantly but never does anything about them? In other words, do you want deeds or talk?

Consider the matter of human slavery. Surely that is a matter that should exercise those who are interested in the spiritual needs of mankind. Is it right, is it just, is it moral, for one man to be slavemaster and another man to be slave? Surely this is not a question for a scientist, since it is not something that can be solved by studying reactions in test tubes or by observing the shifting of needles on the dials of spectrophotometers. The question is for philosophers and theologians, and we all know they have had ample time to consider it.

Throughout the history of civilization, right down to modern times, the wealth and prosperity of a relatively small number of people has been built on the animal-like labor and wretched existence of a large number of peasants, serfs, and slaves. What have our spiritual leaders had to say about it?

In our Western civilization, at least, the prime source of spiritual comfort is the Bible. Look through the Bible, then, from the first verse of Genesis to the last verse of Revelation and you will find not one word of condemnation of slavery as an institution. There are lots of generalizations about love and charity, but no practical suggestions as to governmental responsibility for the poor and unfortunate.

Look through all the writings of the great philosophers of the past and you will find no whisper of condemnation against slavery as an institution. To Aristotle, it seemed quite clear that there were people who seemed to be fitted by temperament to be slaves.

It was, indeed, quite the other way around. Spiritual leaders very often rallied to the support of slavery as an institution, either directly or indirectly. There were not wanting those who justified the forcible abduction of African Blacks into American slavery by saying that they were, in this way, made into Christians and that the salvation of their souls more than made up for the enslavement of their bodies.

Then, too, when religion caters to the spiritual needs of slaves and serfs by assuring them that their Earthly position is God's will and by promising them a life of eternal bliss after death if they do not commit the sin of rebelling against God's will, who is benefited more? Is it the slave whose life may be made more bearable in the contemplation of heaven? Or is it the slavemaster who need be that much less concerned with ameliorating the hard lot of the downtrodden and that much less fearful of revolt?

When, then, did slavery come to be recognized as a grievous and unjustifiable wrong? When did slavery come to an end?

Why, with the dawn of the Industrial Revolution, when machines began to replace muscles.

For that matter, when did democracy on a large scale become possible? When the means of transportation and communication in an industrial age made it possible to work out the mechanics of a representative legislature over wide areas, and when the flood of cheap machine-made goods of all kinds made the "lower classes" into valuable customers who deserved to be coddled.

And what do you suppose would happen if we turned away from science now? What if a noble young generation abandoned the materialism of an industry that seemed to be concerned with things rather than with ideals, and moved, best foot backward, into a world in which everyone moaned and whined about love and charity? Why, without the machinery of our materialistic industry, we would inevitably drift back to a slave economy and we could use love and charity to keep the slaves quiet.

Which is better? Amoral science which puts an end to slavery, or spirituality which in thousands of years of talk didn't?

Nor is slavery the only point we can make.

In the preindustrial age, mankind was subject to the constant onslaught of infectious disease. All the love of parents, all the prayers of congregations, all the lofty generalizations of philosophers could not prevent a child dying of diphtheria or half a nation dying of the plague.

It was the cold curiosity of men of science, working without value judgments, that magnified and studied the forms of life invisible to the unaided eye, that worked out the cause of infectious disease, that demonstrated the importance of hygiene, of clean food and water, of efficient sewage systems. It was that which worked out vaccines, antitoxins, chemical specifics, and antibiotics. It was that which saved hundreds of millions of lives.

It was scientists, too, who won the victory against pain and who discovered how to soothe physical anguish when neither prayer nor philosophy could. There are not many patients facing operations who would demand spiritual solace as a substitute for an anesthetic.

Is it *only* science that is to be praised?

Who can argue against the glories of art, music, and literature that existed long before science did? And what can science offer us to compare with such beauty?

For one thing, it is possible to point out that the vision of the Universe made apparent by the careful labor of four centuries of modern scientists far outweighs in beauty and majesty (for those who would

193

take the trouble to look) all the creations of all human artists put together, or all the imaginings of mythologists, for that matter.

Beyond that, it is also a fact that before the days of modern technology, the full flower of art and the human intellect was reserved for the few who were aristocratic and rich. It was modern science and technology that made books plentiful and cheap. It was modern science and technology that made art, music, and literature available to all and brought the marvels of the human mind and soul to even the meanest.

But haven't science and technology brought us all sorts of undesirable side effects, from the danger of nuclear war to the noise pollution of hard rock on transistor radios?

Yes, and that's nothing new. Every last technological advance, however primitive, has brought with it something undesirable. The stone-tipped axe brought mankind more food—and made war more deadly. The use of fire gave mankind light, warmth, more and better food—and the possibility of arson and of burning at the stake. The development of speech made mankind human—and liars, one and all.

But the choice between good and evil is man's—

In 1847, the Italian chemist Ascanio Sobrero produced nitroglycerine for the first time. He heated a drop of it and it exploded shatteringly. Sobrero realized, in horror, its possible application to warfare and stopped all research in that direction at once.

It didn't help, of course. Others followed up the work and it, along with other high explosives, were being used in warfare within half a century.

Did that make high explosives entirely bad? In 1866, the Swedish inventor Alfred Bernhard Nobel learned how to mix nitroglycerine with diatomaceous earth to produce a mixture that was completely safe to handle and which he called "dynamite." With dynamite, earth could be moved at a rate far beyond the pick-and-shovel efforts of all the ages before, and without brutalizing men at hard labor.

It was dynamite that helped forge the way for the railroads in the final decades of the nineteenth century, that helped construct dams, subways, building foundations, bridges, and a thousand other grand-scale constructions of the industrial age.

It is, after all, mankind's choice whether to use explosives to construct or to destroy. If he chooses the latter, the fault is not in the explosive but in mankind's folly.

Of course, you might argue that all the good that explosives can do isn't worth the harm they can do. You might argue that mankind is incapable of choosing the good and shunning the evil and therefore, as a pack of fools, must be denied explosives altogether.

In that case, let us think back to the medical advances that began with Jenner's discovery of vaccination in 1798, Pasteur's enunciation of the germ theory of disease in the 1860s, and so on. That has doubled man's average life-span, which is good, and has brought on the population explosion, which is bad.

As far as I can see, hardly anyone objects to advances in medicine. Even today, when so many people are concerned about the dangers of scientific and technological advance, I hear hardly any protests against research into the causes and cure of arthritis, circulatory disease, birth defects, or cancer.

And yet the population explosion is the most immediate danger mankind faces. If we avoid nuclear war, counteract pollution, learn to economize on our natural resources, and advance in every field of science, we will nevertheless be destroyed in a matter of decades if the population explosion continues unchecked.

Of all mankind's follies, that of allowing the death rate to drop faster than the birthrate is the worst.

So who's for the abolition of medical advance and return to a high death rate? Who will march under the banner of "Up with Epidemics!"? (Of course, you may consider that epidemics are okay on some other continent—but they have a bad habit of spreading.)

Well, then, shall we pick and choose? Shall we keep

medical advances and a few other noble examples of scienctific progress and abandon the rest of technology? Shall we retire to farms and live in blameless rural splendor, forgetting the wicked city and its machines?

But the farms must have no machinery either—no powered tractors, reapers, binders, and all the rest. They must be without synthetic fertilizers and pesticides, which are the product of an advanced technology. They must be without irrigation machinery, modern dams, and so on. They must be without advanced genetic strains that require plenty of fertilizer and irrigation. —It has to be that way or you've got the entire mechanism of industrialization about your neck again.

In that case, however, world farming can support about one billion people on Earth and there happen to be four billion on Earth right now.

Three billion people, at least, have to be removed from the Earth if we're to become a planet of happy farmers. Any volunteers? No fair volunteering other people; is there anyone who wants to volunteer *himself* for removal. —I thought so.

In the same article, previously cited, in which Toynbee talked about spiritual needs, he also said: "The reason science does succeed in answering its questions is that these quetsions are not the most important ones. Science has not taken up religion's fundamental questions, or, if it has taken these up, it has not given genuine scientific answers to them."

What does Professor Toynbee want? Through advances in science we have ended slavery; brought more security, health, and creature comfort to more people than was dreamed of in all the centuries before science; made art and leisure available to hundreds of millions. And this is as a result of answering questions that "are not the most important ones." Maybe so, Professor, but I am a humble man and these unimportant questions seem pretty good to me if that's what they bring.

And how has religion answered its "fundamental

questions." What are the answers? Is the mass of humanity rather a testimonial to the failure of thousands of years of merely talking about goodness and virtue.

Is there any indication that some particular group of mankind under some particular religion is more moral or more virtuous or more decent than other groups of mankind under other particular religions or, for that matter, under no particular religion—either now or in the past. I have never heard of any such indication. If science could point to no better record of accomplishment than religion can point to, science would have vanished long ago.

The Emperor has no clothes, but superstitious awe seems to prevent the fact from being pointed out.

Let's summarize it, then—

You may not like the route taken by modern science and technology, but there is no other.

Name any world problem and I can tell you that, although science and technology *may not* solve it, anything else *cannot* solve it. So you have the choice: Possible victory with science and technology, or certain defeat without it.

Which do you choose?

15

Thinking About Thinking

I have just returned from a visit to Great Britain (see Chapter 5). In view of my antipathy to traveling (which has not changed), I never thought I would walk the streets of London or stand under the stones of Stonehenge, but I did. Of course, I went by ocean liner both ways, since I don't fly.

The trip was an unqualified success. The weather during the ocean crossing was calm; the ships fed me (alas) all I could eat; the British were impeccably kind to me, even though they did stare a bit at my varicolored clothes, and frequently asked me what my bolo ties were.

Particularly pleasant to me was Steve Odell, who was publicity director of Mensa, the organization of high-IQ people which more or less sponsored my visit. Steve squired me about, showed me the sights, kept me from falling into ditches and under cars, and throughout maintained what he called his "traditional British reserve."

For the most part, I managed to grasp what was said to me despite the funny way the British have of talking. One girl was occasionally incomprehensible, however, and I had to ask her to speak more slowly. She seemed amused by my failure to understand her, although I, of course, attributed it to her imperfect command of the language. "You," I pointed out, "understand *me*."

"Of course I understand you," she said. "You speak slowly in a Yankee drool."

I had surreptitiously wiped my chin before I realized that the poor thing was trying to say "drawl."

But I suppose the most unusual part of the trip (which included three speeches, three receptions, innumerable interviews by the various media, and five hours of book-signing at five bookstores in London and Birmingham) was being made a vice-president of International Mensa.

I took it for granted that the honor was bestowed upon me for the sake of my well-known intelligence, but I thought of it during my five-day return on the *Queen Elizabeth 2* and it dawned on me that I didn't really know much about intelligence. I *assume* I am intelligent, but how can I *know?*

So I think I had better think about it—and where better than here among all my Gentle Friends and Readers?

One common belief connects intelligence with (1) the ready accumulation of items of knowledge, (2) the retention of such items, and (3) the quick recall, on demand, of such items.

The average person, faced with someone like myself (for instance) who displays all these characteristics in abundant degree is quite ready to place the label of "intelligent" upon the displayer and to do so in greater degree the more dramatic the display.

Yet surely this is wrong. One may possess all three characteristics and yet give evidence of being quite stupid; and, on the other hand, one may be quite unremarkable in these respects and yet show unmistakable signs of what would surely be considered intelligence.

During the 1950s, the nation was infested with television programs in which large sums were paid out to those who could come up with obscure items of information on demand (and under pressure). It turned out that some of the shows weren't entirely honest, but that is irrelevant.

Millions of people who watched thought that the

mental calisthenics indicated intelligence.* The most remarkable contestant was a postal employee from St. Louis who, instead of applying his expertise to one category as did others, took the whole world of factual items for his province. He amply displayed his prowess and struck the nation with awe. Indeed, just before the quiz-program fad collapsed, there were plans to pit this man against all comers in a program to be entitled "Beat the Genius."

Genius? Poor man! He had barely competence enough to make a poor living and his knack of total recall was of less use to him than the ability to walk a tightrope would have been.

But not everyone equates the accumulation and ready regurgitation of names, dates, and events with intelligence. Very often, in fact, it is the lack of this very quality that is associated with intelligence. Have you never heard of the absent-minded professor?

According to one kind of popular stereotype, all professors, and all intelligent people generally, are absent-minded and couldn't remember their own names without a supreme effort. But then what makes them intelligent?

I suppose the explanation would be that a very knowledgeable person bends so much of his intellect to his own sector of knowledge that he has little brain to spare for anything else. The absent-minded professor is therefore forgiven all his failings for the sake of his prowess in his chosen field.

Yet that cannot be the whole story either, for we divide categories of knowledge into a hierarchy and reserve our admiration for some only, labeling successful jugglery in those and those only as "intelligent."

We might imagine a young man, for instance, who has an encyclopedic knowledge of the rules of baseball, its procedures, its records, its players, and its cur-

* I was asked to be on one of these shows and refused, feeling that I would gain nothing by a successful display of trivial mental pyrotechnics and would suffer needless humiliation if I were human enough to muff a question.

rent events. He may concentrate so thoroughly on such matters that he is extremely absent-minded with respect to mathematics, English grammar, geography, and history. He is not then forgiven his failure in some respects for the sake of his success in others; he is *stupid!* On the other hand, the mathematical wizard who cannot, even after explanation, tell a bat boy from a home run is, nonetheless, *intelligent*.

Mathematics is somehow associated with intelligence in our judgments and baseball is not, and even moderate success in grasping the former is enough for the label of intelligent, while supreme knowledge of the latter gains you nothing in that direction (though much, perhaps, in others).

So the absent-minded professor, as long as it is only his name he doesn't remember, or what day it is, or whether he has eaten lunch or has an appointment to keep (and you should hear the stories about Norbert Wiener), is still intelligent as long as he learns, remembers, and recalls a great deal about some category *associated* with intelligence.

And what categories are these?

We can eliminate every category in which excellence involves merely muscular effort or co-ordination. However admirable a great baseball player or a great swimmer, painter, sculptor, flautist, or cellist may be, however successful, famous, and beloved, excellence in these fields is, in itself, no indication of intelligence.

Rather it is in the category of theory that we find an association with intelligence. To study the technique of carpentry and write a book on the various fashions of carpentry through the ages is a sure way of demonstrating intelligence even though one could not, on any single occasion, drive a nail into a beam without smashing one's thumb.

And if we confine ourselves to the realm of thought, it is clear that we are readier to associate intelligence with some fields than with others. We are almost sure to show more respect for a historian than for a sports writer, for a philosopher than for a cartoonist, and so on.

It seems an unavoidable conclusion to me that our notions of intelligence are a direct inheritance from the days of ancient Greece, when the mechanical arts were despised as fit for artisans and slaves, while only the "liberal" arts (from the Latin word for "free men") were respectable, because they had no practical use and were therefore fit for free men.

So nonobjective is our judgment of intelligence, that we can see its measure change before our eyes. Until fairly recently, the proper education for young gentlemen consisted very largely in the brute inculcation (through beatings, if necessary) of the great Latin writers. To know no Latin seriously disqualified anyone for enlistment in the ranks of the intelligent.

We might, of course, point out that there is a difference between "educated" and "intelligent" and that the foolish spouting of Latin marked only a fool after all—but that's just theory. In actual fact, the uneducated intelligent man is invariably downgraded and underestimated and, at best, is given credit for "native wit" or "shrewd common sense." And women, who were not educated, were shown to be unintelligent by their lack of Latin and that was the excuse for not educating them. (Of course that's circular reasoning, but circular reasoning has been used to support all the great injustices of history.)

Yet see how things change. It used to be Latin that was the mark of intelligence and now it is science, and I am the beneficiary, I know no Latin except for what my flypaper mind has managed to pick up accidentally, but I know a great deal of science—so without changing a single brain cell, I would be dumb in 1775 and terribly smart in 1975.

You might say that it isn't knowledge itself, not even the properly fashionable category of knowledge, that counts, but the *use* that is made of it. It is, you might argue, the manner in which the knowledge is displayed and handled, the wit, originality, and creativity with which it is put to use, that counts. Surely, *there* is the measure of intelligence.

And to be sure, though teaching, writing, scientific research are examples of professions often associated with intelligence, we all know there can be pretty dumb teachers, writers, and researchers. The creativity or, if you like, the intelligence can be missing and still leave behind a kind of mechanical competence.

But if creativity is what counts, that, too, only counts in approved and fashionable areas. A musician, unlearned, uneducated, unable to read music perhaps, may be able to put together notes and tempos in such a way as to create, brilliantly, a whole new school of music. Yet that in itself will not earn him the accolade of "intelligent." He is merely one of those unaccountable "creative geniuses" with a "gift from God." Since he doesn't know how he does it, and cannot explain it after he's done it,† how can he be considered intelligent?

The critic who, after the fact, studies the music, and finally, with an effort, decides it is not merely an unpleasant noise by the old rules, but is a great accomplishment by certain new rules—why he *is* intelligent. (But how many critics would you exchange for one Louis Armstrong?)

But in that case, why is the brilliant scientific genius considered intelligent? Do you suppose he knows how his theories come to him or can explain to you how it all happened? Can the great writer explain how he writes so that you can do as he does?

I am not, myself, a great writer by any standard I respect, but I have my points and I have this value for the present occasion—that I am one person, generally accepted as intelligent, whom I can view from within.

---

† The great trumpeter Louis Armstrong, on being asked to explain something about jazz, is reported to have said (translated into conventional English), "If you've got to ask, you aren't ever going to know." —These are words fit to be inscribed on jade in letters of gold.

Well, my clearest and most visible claim to intelligence is the nature of my writing—the fact that I write a great many books in a great many fields in complex yet clear prose, displaying great mastery of much knowledge in doing so.

So what?

No one ever taught me to write. I had worked out the basic art of writing when I was eleven. And I can certainly never explain what that basic art is to anyone else.

I dare say that some critic, who knows far more of literary theory than I do (or than I would ever care to), might, if he chose, analyze my work and explain what I do and why, far better than I ever could. Would that make him more intelligent than I am? I suspect it might, to many people.

In short, I don't know of any way of defining intelligence that does not depend on the subjective and the fashionable.

Now, then, we come to the matter of intelligence-testing, the determination of the "intelligence quotient" or "IQ."

If, as I maintain and firmly believe, there is no objective definition of intelligence, and what we call intelligence is only a creation of cultural fashion and subjective prejudice, what the devil is it we test when we make use of an intelligence test?

I hate to knock the intelligence test, because I am a beneficiary of it. I routinely end up on the far side of 160 when I am tested and even then I am invariably underestimated because it almost always takes me less time to do a test than the time allotted.

In fact, out of curiosity, I got a paperback book containing a sizable number of different tests designed to measure one's IQ. Each test had a half-hour time limit. I worked on each one as honestly as I could, answering some questions instantly, some after a bit of thought, some by guesswork, and some not at all.—And naturally, I got some answers wrong.

When I was done, I worked out the results according to directions and it turned out I had an IQ of 135. — But wait! I had not accepted the half-hour limit offered me, but broke off each section of the test at the fifteen-minute mark and went on to the rest. I therefore doubled the score and decided I have an IQ of 270. (I'm sure that the doubling is unjustified, but the figure of 270 pleases my sense of cheerful self-appreciation, so I intend to insist on it.)

But however much all this soothes my vanity, and however much I appreciate being vice-president of Mensa, an organization which bases admission to its membership on IQ, I must, in all honesty, maintain that it means nothing.

What, after all, does such an intelligence test measure but those skills that are associated with intelligence by the individuals designing the test? And those individuals are subject to the cultural pressures and prejudices that force a subjective definition of intelligence.

Thus, important parts of any intelligence test measure the size of one's vocabulary, but the words one must define are just those words one is apt to find in reading approved works of literature. No one asks for the definition of "two-bagger" or "snake eyes" or "riff," for the simple reason that those who design the tests don't know these terms or are rather ashamed of themselves if they do.

This is similarly true of tests of mathematical knowledge, of logic, of shape-visualization, and of all the rest. You are tested in what is culturally fashionable—in what educated men consider to be the criteria of intelligence—i.e., of minds like their own.

The whole thing is a self-perpetuating device. Men in intellectual control of a dominating section of society define themselves as intelligent, then design tests that are a series of clever little doors that can let through only minds like their own, thus giving them more evidence of "intelligence" and more examples of "intelligent people" and therefore more reason to devise additional tests of the same kind. More circular reasoning!

And once someone is stamped with the label "Intelligent" on the basis of such tests and such criteria, any demonstration of stupidity no longer counts. It is the label that matters, not the fact. I don't like to libel others, so I will merely give you two examples of clear stupidity which I myself perpetrated (though I can give you two hundred, if you like)—

1) On a certain Sunday, something went wrong with my car and I was helpless. Fortunately, my younger brother, Stan, lived nearby and since he is notoriously goodhearted, I called him. He came out at once, absorbed the situation, and began to use the yellow pages and the telephone to try to reach a service station, while I stood by with my lower jaw hanging loose. Finally, after a period of strenuous futility, Stan said to me with just a touch of annoyance, "With all your intelligence, Isaac, how is it you lack the brains to join the AAA?" Whereupon, I said, "Oh, I belong to the AAA," and produced the card. He gave me a long, strange look and called the AAA. I was on my wheels in half an hour.

2) Sitting in Ben Bova's room (he's editor of *Analog*) at a recent science fiction convention, I was waiting, rather impatiently, for my wife to join us. Finally, there was a ring at the door. I sprang to my feet with an excited "Here's Janet!" flung open a door, and dashed into the closet—when Ben opened the room door and let her in.

Stan and Ben love to tell these stories about me and they're harmless. Because I have the label "intelligent," what would surely be evidence of stupidity is converted into lovable eccentricity.

This brings us to a serious point. There has been talk in recent years of racial differences in IQ. Men like William B. Shockley, who has a Nobel Prize (in physics), point out that measurements show the average IQ of Blacks to be substantially lower than that of Whites, and this has created quite a stir.

Many people who, for one reason or another, have already concluded that Blacks are "inferior" are delighted to have "scientific" reason to suppose that the undesirable position in which Blacks find themselves is their own fault after all.

Shockley, of course, denies racial prejudice (sincerely, I'm sure) and points out that we can't deal intelligently with racial problems if, out of political motives, we ignore an undoubted scientific finding; that we ought to investigate the matter carefully and study the intellectual inequality of man. Nor is it just a matter of Blacks versus Whites; apparently some groups of Whites score less well than do other groups of Whites, and so on.

Yet to my mind the whole hip-hurrah is a colossal fraud. Since intelligence is (as I believe) a matter of subjective definition and since the dominant intellectuals of the dominant sector of society have naturally defined it in a self-serving manner, what is it we say when we say that Blacks have a lower average IQ than Whites have? What we are saying is that the Black subculture is substantially different from the dominant White subculture and that the Black values are sufficiently different from dominant White values to make Blacks do less well on the carefully designed intelligence tests produced by the Whites.

In order for Blacks, on the whole, to do as well as Whites, they must abandon their own subculture for the White and produce a closer fit to the IQ-testing situation. This they may not want to do; and even if they want to, conditions are such that it is not made easy for them to fulfill that desire.

To put it as succinctly as possible: Blacks in America have had a subculture created for them, chiefly by White action, and have been kept in it chiefly by White action. The values of that subculture are defined as inferior to those of the dominant culture, so that the Black IQ is arranged to be lower; and the lower IQ is then used as an excuse for the continuation of the very conditions that produced it. Circular reasoning? Of course.

But then, I don't want to be an intellectual tyrant and insist that what I speak must be the truth.

Let us say that I am wrong; that there *is* an objective definition of intelligence, that it *can* be measured accurately, and that Blacks *do* have lower IQ ratings than Whites do, on the average, not because of any cultural differences but because of some innate, biologically based intellectual inferiority. Now what? How should Whites treat Blacks?

That's a hard question to answer, but perhaps we can get some good out of supposing the reverse. What if we test Blacks and find out, more or less to our astonishment, that they end up showing a *higher* IQ than do Whites, on the average?

How should we *then* treat them? Should we give them a double vote? Give them preferential treatment in jobs, particularly in the government? Let them have the best seats in the bus and theater? Give them cleaner restrooms than Whites have, and a higher average pay scale?

I am *quite* certain that the answer would be a decided, forceful, and profane negative for each of these propositions and any like them. I suspect that if it were reported that Blacks had higher IQ ratings than Whites do, most Whites would at once maintain, with considerable heat, that IQ could not be measured accurately and that it was of no significance if it could be, that a person was a person regardless of book learning, fancy education, big words, and fol-de-rol, that plain ordinary horsesense was all anyone needed, that all men were equal in the good old United States, and those damned pinko professors and their IQ tests could just shove it—

Well, if we're going to ignore IQ when *we* are on the low end of the scale, why should we pay such pious attention to it when *they* are?

But hold on. I may be wrong again. How do I know how the dominants would react to a high-IQ minority? After all, we *do* respect intellectuals and professors to a certain extent, don't we? Then, too, we're talking about oppressed minorities, and a high IQ minority wouldn't be oppressed in the first place, so the artificial situation

I set up by pretending the Blacks scored high is just a straw man, and knocking it down has no value.

Really? Let's consider the Jews, who, for some two millennia, have been kicked around whenever Gentiles found life growing dull. Is this because Jews, as a group, are low-IQ? —You know, I *never* heard that maintained by anyone, however anti-Semitic.

I do not, myself, consider Jews, as a group, to be markedly high-IQ. The number of stupid Jews I have met in the course of a lifetime is enormous. That, however, is not the opinion of the anti-Semite, whose stereotype of the Jews involves their possession of a gigantic and dangerous intelligence. Although they may make up less than half a percent of a nation's population, they are forever on the point of "taking over."

But then, shouldn't they, if they are high-IQ? Oh, no, for that intelligence is merely "shrewdness," or "low cunning," or "devious slyness," and what really counts is that they lack the Christian, or the Nordic, or the Teutonic, or the what-have-you virtues of other sorts.

In short, if you are on the receiving end of the game-of-power, any excuse will do to keep you there. If you are seen as low-IQ, you are despised and kept there because of that. If you are seen as high-IQ, you are feared and kept there because of that.

Whatever significance IQ may have, then, it is, at present, being made a game for bigots.

Let me end, then, by giving you my own view. Each of us is part of any number of groups corresponding to any number of ways of subdividing mankind. In each of these ways, a given individual may be superior to others in the group, or inferior, or either, or both, depending on definition and on circumstance.

Because of this, "superior" and "inferior" have no useful meaning. What *does* exist, objectively, is "different." Each of us is different. I am different, and you are different, and you, and you, and you—

It is this difference that is the glory of *Homo sapiens* and the best possible salvation, because what some cannot do, others can, and where some cannot flourish,

209

others can, through a wide range of conditions. I think we should value these differences as mankind's chief asset, as a species, and try never to use them to make our lives miserable, as individuals.

16

Star in the East

Because I am an occasional writer of light verse, and a punster, and also an egocentric, I am sometimes compelled to do something clever (if I can) with my name. Thus, I once wrote a poem, "The Prime of Life," in which I needed an internal rhyme and I wanted to use my name, so I had some young fan meet me and say:

"Why, stars above, it's Asimov"

I thought that was a natural, unforced line and I quoted it sometimes when I wanted to impress someone with my skill at light verse. I did so once to a fair damsel and she gave it some five seconds of thought and said, "Why didn't you say:

'Why, mazel tov, it's Asimov'?"

It took me some fifteen minutes of speechless chagrin before I recovered. Her version was much better of course, for "mazel tov" (as I perhaps need not tell you) is the Hebrew phrase for "good luck." It is much more humorously appropriate for several reasons—and I had never thought of it.

The cleverest use of my name, however, was not by myself, but by J. Wayne Sadler of Jacksonville, Florida. Last December, he sent me a verse (into which I have introduced two or three tiny changes) and here it is—

When Isaac's at a nudist camp
He promptly joins the fun,

For "When in Rome" 's his favorite quote
  As he tells everyone.
So when the signal's given out,
  "All clothing you must doff,"
Without a moment's hesitation,
  Isaac Asimov.

Ah, well, I've never been in a nudist camp, but I often feel that, thanks to my personal style of writing, I live in a mental nudist camp. There can be no one who reads me regularly who isn't completely aware of my opinions and feelings on almost any subject. Still, let me state, if you can possibly have missed it, that I am a freethinker in religion.

In particular, I must explain that I do not accept as accurate the nativity tales that appear in the Gospels. As to their theological value, or their allegorical symbolism, or what not, I have nothing to say; I am not a theologian. I do not accept them, however, as portrayals of the literal truth, any more than I accept Genesis 1.

My own feeling is that the tales of the nativity were devised after the fact, and in many ways follow the tradition of the nativity tales that were told of earlier legendary (or not-so-legendary) leaders who founded nations or religions: Sargon of Agade, Moses, Cyrus, Romulus and Remus, and so on.

The oldest of the four Gospels, Mark, contains no nativity tale at all, but begins with the baptism of a mature Jesus. The youngest of the four Gospels, John, has no human nativity tale because Jesus had, in a way, gotten beyond that by then. Instead it contains a treatment of Jesus as a manifestation of God and as coeternal with him.

That leaves us with two Gospels of intermediate age, Matthew and Luke, each of which contains a nativity tale—but a different one. The two do not overlap at any single point: anything contained in one nativity tale is omitted in the other.

Thus the story of the star that shone in the sky at the time of the birth of Jesus, is found only in the Gospel of Matthew and does *not* occur in any form in the Gospel of Luke. Indeed, the star is not referred to anywhere in the New Testament but in the first part of the second chapter of Matthew.

The whole story of this star is to be found in five verses, and here they are as the King James version has it:

Now when Jesus was born in Bethlehem of Judaea in the days of Herod the king, behold, there came wise men from the east to Jerusalem, Saying, Where is he that is born King of the Jews? for we have seen his star in the east, and are come to worship him. [Matthew 2:1–2]

This interests King Herod, who wants no pretender to the throne and who would naturally expect any so-called Messiah to stir up rebellions. He summons his advisers, then sends for the wise men.

Then Herod, when he had privily called the wise men, enquired of them diligently what time the star appeared. [2:7]

Herod then instructs the wise men to find the child and report back to him.

When they had heard the king, they departed; and, lo, the star, which they saw in the east, went before them, till it came and stood over where the young child was. When they saw the star, they rejoiced with exceeding great joy. [2:9–10]

Because this star shone above the place of Jesus' birth in Bethlehem (wherever in the town that might have been, for the story of the manger is found only in

Luke), it is usually referred to as "the Star of Bethlehem."

The Star of Bethlehem is one of the few matters in the Bible that seems to be astronomical in nature, and it has therefore been the subject of a great deal of speculation from the astronomical viewpoint. And, to tell you the truth, speculating about the Star of Bethlehem is my kind of game, too, so I would like to present you with no less than nine alternatives.

It might be, for instance *(alternative 1),* that the Star of Bethlehem is not amenable to any astronomical explanation, and rests, indeed, outside the realm of reason altogether. It may represent a "mystery" (in the religious sense of the word) that human beings cannot understand without divine inspiration. Perhaps only in Heaven can the full meaning dawn. In that case, clearly, there is no use in speculating. We can do nothing but wait for inspiration or for arrival in Heaven, and, alas, neither is likely to happen to me.

It might also be *(alternative 2)* that the Star of Bethlehem is beyond explanation, not for theological reasons, but simply because it is a pious invention on the part of the writer of the Gospel.

This is not to say that it is a deliberate lie or a purposeful attempt to deceive. The tale of the star may have been in the air, as one of the standard indications of divinity at birth—just as angelic voices and halos might be—and the author made use of it, as a suitable and fitting detail.

Remember that Matthew was probably putting together his Gospel some time after the destruction of the Temple in A.D. 70; in other words, three quarters of a century after the birth of Jesus. There were no records of the past in the modern sense and he could only gather vague tales. There may have been stories of some starlike phenomenon having taken place at about the time of Jesus' birth and Matthew felt it would be appropriate to include it.

We might ask why Matthew was impressed by the

tales of the star which he heard, and wanted to include it, when Luke didn't. Actually, we can advance a plausible reason for this. From external evidence, we can argue that Luke was a Gentile, telling the gospel story to Gentiles, while Matthew was a Jew telling the story to Jews.*

It is natural, then, for Matthew to present as many details as possible that bear out some Old Testament prophecy or other since this would impress his Jewish audience. Sometimes he cites the Old Testament verses that contain the prophecy, but even when he doesn't we might look for one.

At one point in the Old Testament, for instance, Balaam is described as making the following prophecy at the time that the Israelite tribes are preparing, east of the Jordan, to invade Canaan:

> I shall see him, but not now. I shall behold him, but not nigh: there shall come a Star out of Jacob, and a Sceptre shall rise out of Israel, and shall smite the corners of Moab, and destroy all the children of Sheth. [Numbers 24:17]

It is very likely that this verse was written during the time of the Judean kingdom and was included as part of the words of the legendary sage Balaam. (It was common in ancient times to place words in the mouths of ancient worthies.)

The assumption is that the "him" is King David, who did defeat Moab and conquer all the surrounding kingdoms. It is because of this verse that the two interlocked equilateral triangles are referred to as "the Star of David."

After the destruction of the kingdom of Judah and the end of the Davidic dynasty, the verse was reinterpreted. The word "him" was taken as referred to a future king of the Davidic dynasty, the Messiah ("anointed

---

* You are welcome to refer to my book *Asimov's Guide to the Bible*, vol. 2, *The New Testament* (Doubleday, 1969) if you wish. I don't insist.

215

one," a phrase commonly used by the Jews to refer to a king). Matthew naturally accepted it as such and would suppose that a star would make a particularly fit association with the birth of the Messiah.

Then, too, there is a passage in Isaiah that describes a utopia to come. One verse goes:

And the Gentiles shall come to thy light, and kings to the brightness of thy rising. [Isaiah 60:3]

The reference is to the ideal Israel that is to rise in the future, but it is easy to transfer that reference to the Messiah, and the words "light" and "brightness of thy rising" could refer to a star. The word "Gentiles" may be taken to refer to the wise men from the east.

So influential is the Isaiah verse, with its reference to "kings" as well as "Gentiles," that the legend arose that the wise men were three kings named Melchior, Gaspar, and Balthazar. In medieval times, relics of the three were supposed to exist at the Cologne Cathedral, so that they came to be called "the Three Kings of Cologne." All this is, of course, quite nonbiblical. The Bible does not call them kings and does not even say there were three of them.

But what if Matthew did base the tale of the star on some legend current at the time the Gospel was being put together, and what if the legend reflected something that had actually happened?

We might (alternative 3) suppose that whatever the star was, it was a miraculous object and not something that could be seen in the ordinary course of events or by everybody. It might, in fact, have been visible only to the wise men and have indeed served as their miraculous guide. Once it had reached the infant Jesus and stood over him, it disappeared.

We might argue in favor of this by pointing out that Herod, who would be expected to be keenly interested in any sign that might indicate the birth of a rival to his throne, knew nothing about the star and had to inquire of the wise men concerning it.

But if the star is a miracle created for this one task and seen by only the people who had to see it, further investigation must stop here, so let's proceed to further alternatives.

Let us suppose that the star, whatever it was, was not miraculous, but was real, and was something that could be visible to anyone who looked. This, indeed, is the assumption that most people make when they try to work out what the Star of Bethlehem might have been.

In any alternative arising from this assumption, however, we must forget about the star guiding the wise men and standing over Jesus. That is clearly miraculous and must be omitted if a rational explanation is sought. We must simply suppose that something appeared in the sky which seemed to betoken the birth, of a Messiah, and no more than that.

We are helped in this however, by the fact that the term "star" had much wider meaning to the ancients than to ourselves. We consider planets and comets to be nonstars, for instance, but to the ancients they were "wandering stars" and "hairy stars" respectively. To the ancients, any heavenly object would be considered a star, so let's search for one in the broadest possible way.

For instance, the heavenly phenomenon referred to as a star by Matthew may actually have been *(alternative 4)* a subtle astronomical fact that was real enough, but would not be apparent to anyone but specialists in the field.

The wise men might well have been considered specialists in the field. The term as used in Matthew is an English translation of the Greek word *magoi* which is, in turn, from *magu,* the name given by the ancient Persians to Zoroastrian priests.

To the Greeks and Romans, the term referred to any eastern mystic. To the Romans, *magus* (plural *magi)* came to mean "sorcerer" and, as a matter of fact, our present English words "magic" and "magician" come from Persian *magu.*

The most likely people to be interested in heavenly

phenomena were, of course, astrologers, and they would come under the heading of magi. Babylonia was an ancient center of astrology, so the wise men might easily have been astrologers from that land, which lies east of Judea.

And what could the astrologers have seen that was apparent to them, and real, but which ordinary people could not see?

It so happens that the position of the Sun at the time of the vernal equinox is of importance to astrologers. This position is always in the Zodiac, but is not fixed. It very slowly shifts from one of the twelve constellations of the Zodiac to the next, taking about two thousand years to pass completely through one constellation.†

For the two thousand years prior to the birth of Jesus, the Sun at the time of the vernal equinox had been in the constellation of Aries (the Ram). Now, however, it was more or less on the point of moving into the constellation of Pisces (the Fish). To astrologers this would be a most vital event and might well be thought to represent some basic upset in human events. Since the Judeans of the time were constantly talking of the arrival of a Messiah, who would establish a new Jerusalem and reorganize human history (as in the Isaiah passage), astrologers might conclude that would be it—and go to Judea to investigate the matter.

It is interesting this connection that the early Christians used a fish as a secret symbol of the Messiah. The usual explanation is that the letters of the Greek word for "fish," taken in order, were initials of a Greek phrase which, translated, is "Jesus Christ, Son of God, Savior." Yet it might also be that the fish referred to Pisces, into which the vernal equinox had now passed.

Yet the point of vernal equinox is not visible; it is only calculated. Matthew clearly speaks of a visible star. To be sure, this might only be because Matthew, no astrologer, misunderstood what it was all about. Still,

† See "Signs of the Times" in *Of Matters Great and Small* (Doubleday, 1975).

218

we can't know that. Suppose we decide that Matthew was right and that the star was a visible phenomenon. What then?

The star might, in that case *(alternative 5)*, have been a comet. Comets appear irregularly and unpredictably (at least as far as the ancients were concerned) and move erratically across the sky. It so happens that the most famous of them all, Halley's Comet, was in the sky in 11 B.C., which is seven years before the traditional date given for the birth of Jesus, but that traditional date rests on shaky ground.

Yet Halley's Comet is *too* noticeable. Comets are visible to all and were generally taken to portend world-shaking events. If the wise men came from the east talking of a star representing the birth of a Messiah, everyone would know at once what they were talking about and Herod wouldn't have to inquire what it was all about.

The same objection might be raised, less strongly, to *(alternative 6)* the presence of a supernova in the sky, one shining brightly in a position no star had ever before occupied, and therefore signifying something great and new. It would not have been as noticeable as a comet, as far as the general population was concerned, but it is not likely to have gone without comment altogether, and we have no record anywhere of any supernova appearing anywhere around that time of history, nor any trace in the sky today that one might then have appeared.‡

Failing a comet or a supernova, the star might have been *(alternative 7)* a reference to the brightest normal object in the sky, next to the Sun and Moon—the planet Venus. This seems, however, in the highest degree unlikely, although some people have maintained it as a possibility. After all, Venus is a common object in the

‡ The most interesting aspect of this alternative to science fiction readers is that Arthur C. Clarke wrote a story, "The Star," which appeared in the November 1955 issue of *Infinity Science Fiction* and which was awarded a Hugo in 1956. It is the Star of Bethlehem story and, if you don't believe me, I urge you to read it in my anthology *The Hugo Winners* (Doubleday, 1962).

sky and there is no way that it could reasonably be expected to indicate something special at one time rather than another. The same is true to an even greater extent of any other single planet or star in the sky.

What about *(alternative 8)* a bright meteorite? That has an advantage over a comet, a supernova, or a bright planet in that it is a restricted phenomenon. It is located in the upper atmosphere and can be seen only over a very limited portion of the Earth's surface.

The wise men could have seen the "star" in the east, as they announced, in the sky of their own Babylonian area. It would not have been visible elsewhere, and particularly not in Judea. We could then understand how it was that Herod had to inquire about it.

The difficulty here is whether a simple meteorite coming and going would strike astrologers as sufficiently unusual in itself to indicate the coming of a Messiah. In the clear air of Babylonia, there are undoubtedly meteors to be seen every night, and even if this one were particularly bright, so what? If the meteor had actually reached Earth's surface and become a meteorite, the wise men would have been more impressed, provided they had witnessed the fall and found the meteorite, but then would they not have talked of something falling from heaven?

We have run out of the ordinary celestial phenomena that could account for the star—stars themselves, planets, comets, meteors. What is left?

Perhaps it was not a single heavenly object, but a collection of them, an unusual collection that *(alternative 9)* would attract the eyes of astrologers and have significance to them.*

The only objects in the sky that regularly change their position and that form impressive combinations now and then, are the members of the Solar System. Of these, we can omit the comets and meteors, since the former are impressive in themselves and require no combina-

---

* For the data I cite in connection with alternative 9, I am indebted to an article, "Thoughts on the Star of Bethlehem," by Roger W. Sinnott in the December 1968 *Sky and Telescope*.

.ions, and the latter move too rapidly and endure too briefly to form definite combinations. We can omit the Sun, since that drowns out everything in its vicinity and forms no visible combinations; and the Moon as well, since it overwhelms other objects with which it might form a visible combination.

That leaves us the five visible planets—Mercury, Venus, Mars, Jupiter, and Saturn. Every once in a while, two or more of these planets shine close to each other in the sky and very often this makes for a startling combination. Such a situation is not at all uncommon and, according to Sinnott, there were, between 12 B.C. and A.D. 7, no fewer than two hundred occasions when two planets were fairly close to each other in the sky, and twenty occasions when more than two were.

That averages out to roughly one a month and it seems to me that astrologers wouldn't be impressed by them, unless they represented something very unusual, or noticeable, or astrologically significant, or, best of all, all three.

We might set up some criteria. The two brightest planets are Venus and Jupiter. Therefore when those two are close together it is the most dazzling of the combinations, especially when they are sufficiently far from the Sun to be seen in a dark sky.

One such combination took place in the hours before dawn on August 12, 3 B.C. At their closest approach, the two planets were separated by only twelve minutes of arc, a distance only two-fifths the Moon's diameter.

Another similar, but much more striking, combination took place after sunset on June 17, 2 B.C. Venus and Jupiter approached more closely on this occasion and, at the closest, were separated by only three minutes of arc, one-tenth the width of the full Moon.

With that close an approach, it would be difficult to make out the planets as two separate points of light. What's more, as seen from Babylonia, the two planets would be approaching each other steadily as they sank toward the western horizon. Indeed, they would reach their minimum separation at 10 P.M. Babylonian time, just as they were setting. We might imagine that the watching astrologers would see the two planets apparently join into one as they reached a point on the western horizon in the direction of Judea.

Is the fact that the unusual "star" was seen in the direction of Judea enough to make them think of a Messiah? Well, there's more.

An important Messianic prophecy in the Bible is ascribed to Jacob on his death bed. He says something mystical about each of his sons and this is taken as a reference to the future of each tribe.

Concerning Judah (from whom David, and therefore, Jesus, were descended), he said:

> Judah is a lion's whelp: from the prey, my son, thou art gone up: he stooped down, he couched as a lion, and as an old lion; who shall rouse him up? The sceptre shall not depart from Judah, nor a lawgiver from between his feet, until Shiloh come; and unto him shall the gathering of the people be. [Genesis 49:9-10]

The ninth verse is an indication that the lion was the totemic symbol of the tribe of Judah (we still speak of the "Lion of Judah"). As for the tenth verse, there is a lot of argument about the meaning of Shiloh.

Shiloh was the name of a town at which an important shrine existed before the days of the kingdom, and it was destroyed a century before David's time. The verse seems to make little sense in that case and it may be the result of a copyist's error. However, one might argue that it referred to the coming *once more* of the destroyed shrine at Shiloh; hence, by analogy, to the revival of the destroyed Davidic dynasty, and, therefore, to the Messiah. The verse is very commonly viewed as a messianic prophecy.

Now it so happens that one of the constellations of the Zodiac is Leo (the Lion). It would be easy for astrologers to suppose that Leo represents Judah and the House of David. There is a reference to "a lawgiver

222

from between his feet" and between the forefeet of the constellation Leo (as conventionally drawn in ancient times) was its brightest star, Regulus (Latin for "little king"). We might suppose, then, that Regulus, in particular, would represent the Messiah (to astrologers).

As it happens, the Venus-Jupiter combinations of 3 B.C. and of 2 B.C. both took place in the constellation Leo, one on one side of Regulus, and one on the other. In each case, the planetary combination was about three degrees away from Regulus, close enough to be impressive to astrologers.

We have then a single unusual "star" appearing on the horizon over Judea, close to the star that symbolizes the Messiah. Wouldn't you think the astrologers would leave for Judea at once to search for him, even if only to check their own conclusions?

Of course both combinations took place in the summer months and nowhere near the time of Christmas, but that doesn't matter. The December 25 date has no biblical warrant and was chosen in early Christian times merely to compete with the Mithraist festival on that day and to take advantage of the long-established tradition of general jubilation at the time of the winter solstice.

Then, too, both Matthew and Luke place the birth of Jesus in the time of Herod, and that monarch died in 4 B.C. It would seem then that Jesus could not have been born later than 4 B.C. and so could not have been less than two years old at the time of the second and more impressive combination. However, the fact that Jesus was born *precisely* at the time of the appearance of the combination may have been a later improvement on the story.

I must admit that the development of alternative 9 is so attractive that I am tempted to believe it—but I won't. In 2 B.C., astronomy was in the doldrums and even if Babylonian astrologers noted the combination, I doubt that they were so versed in the details of the scriptures and legends of the Judeans as to attach messianic importance to it. No, the whole tale is but an ingenious working-out after the fact.

So I'll stick to my skepticism and place the Star of Bethlehem in the same category with the parting of the Red Sea and with walking on water and with all the other miracles in the Bible. They are merely wonder tales that we would utterly ignore as unworthy of attention, except for the fact that they are *our* wonder tales, which we were taught in impressionable youth to revere.

17

The Judo Argument

In the course of the decades during which I have been explaining the workings of the Universe, without referring to God, I have naturally been asked over and over again whether I believe in God. This is moderately annoying, and I have tried a number of different ways of answering the question, hoping to give no grounds for either argument or offense. (Once, on television, when asked "Do you believe in God?" I answered, "Whose?")

But "belief" doesn't matter anyway, one way or the other. All the hundreds of millions of people who, in their time, believed the Earth was flat never succeeded in unrounding it by an inch.

What we want is some logical line of reasoning, preferably one that starts with observed facts, that leads us to the inescapable conclusion that God exists.

Perhaps that is not possible. Perhaps God's existence is a matter that lies fundamentally beyond the ability of man to observe, measure, and reason out; and must be based on revelation and faith alone. This, in fact, is the attitude of almost all the Believers in our Western culture. They wave the Bible (or some equivalent authority) and that ends the argument.

There's no point in arguing with that, of course. You cannot very well reason with someone whose basic line of argument is that reason doesn't count.

But you know, finding refuge in authority is not necessarily the whole answer. There is a long and respectable series of attempts on the part of impeccably pious people to show that reason does *not* conflict with faith, and that one can begin from first principles and prove by good logic that God exists.

Here, for instance, is a very simple argument for the existence of God. It is called the "ontological argument" ("ontology" being the study of real existence) and was advanced by Saint Anselm in 1078. The argument is that anyone can conceive of a perfect being, which we can call God. But to be truly perfect, such a being must also exist, for nonexistence would be a flaw in perfection. The statement "God does not exist" is, of necessity, a contradiction in terms for it is another way of saying, "The perfect is not perfect." Therefore, God exists.

Not being a theologian, I don't know the proper way of refuting that argument. My own way of refuting it, undoubtedly improper, is to say that as a science fiction writer I daily conceive of things that do not exist and that even to conceive a perfect entity (such as a perfect gas or a perfect black body) does *not* necessarily imply existence.

As far as I know, there is no rational argument designed to prove the existence of God that has been accepted by philosophers and theologians generally. All the arguments remain in dispute and, for complete safety, Believers must fall back on faith.

There is, however, a certain class of argument for the existence of God that particularly interests me, and that is the argument based on science.

After all, ever since the time of Copernicus and Galileo, there has been a general feeling that science and religion are in conflict, and, indeed, many doctrines accepted by science have been bitterly denounced by Believers. The most prominent of these today is the doctrine of evolution by natural selection, with its corollary that life began and developed as a result of natural forces acting in a random way.

When Believers base an argument for the existence of God on scientific findings, they are calling upon the enemy, so to speak. It is a form of philosophical judo—

the art of using the opponent's own strength against him. If you don't mind, then, I will call arguments in favor of the existence of God that are based on scientific findings "judo arguments."

The first judo argument I know of dates back to about 1773, when the French encyclopedist Denis Diderot was at the court of Catherine the Great of Russia. Diderot was an open atheist who expressed his views freely. Leonhard Euler, a Swiss mathematician and one of the greatest of all time, undertook to confound Diderot by proving the existence of God mathematically in open debate.

Diderot accepted the challenge, and with the Russian court looking on in interest, Euler said, "Sir, $(a+b_n)/n=x$, therefore God exists. Refute that!"

Diderot, who knew no mathematics, had no answer, retired in confusion, and asked permission to return to France.

Euler's argument was, of course, nonsense. It was nothing but a practical joke. To this day, there is no mathematical proof of God's existence that anyone of importance accepts.

Let's go on to more serious judo arguments.

Here is one that can be expressed as follows—Suppose something exists, but that it could come into existence only by defying a well-established and universally accepted natural law. We can then argue that the fact of its existence transcends natural law. Since the only factor that has ever been admitted, in our Western culture, to transcend natural law is God, we conclude that God exists.

Examples of this argument have turned up in my mail recently (and not unexpectedly) as a result of my *F & SF* article "Look Long upon a Monkey."* Several people objected to my acceptance of evolution, insisting that life could not evolve through random processes of nature, because "it is impossible to have order arise from

* Included in *Of Matters Great and Small* (Doubleday, 1975).

disorder." The more sophisticated of them said something more formidable: "The concept of evolution violates the second law of thermodynamics."

To be sure, the second law of thermodynamics *does* imply that the quantity of disorder (or "entropy") in the Universe is constantly increasing and that in any spontaneous event it must increase. What's more, no scientist seriously questions the second law of thermodynamics and if any scientific finding can be shown to violate it, that finding is very likely to be thrown out forthwith.

It is further clear that the course of evolution from simple compounds to complex compounds to simple organisms to complex organisms, represents a vast increase of order, or a vast decrease in entropy.

Combining what I have said in the previous two paragraphs, have I not stated that evolution violates the second law of thermodynamics and that, therefore, God exists?

Oddly enough, I haven't. The second law of thermodynamics applies to a "closed system," one that is completely isolated from the rest of the Universe and that neither gains nor loses energy in any form. It is possible to imagine a perfectly closed system and work out the theoretical consequences of the second law, or to construct an almost closed system and observe actual consequences that approach the theoretical ones.

The only true closed system, however, is the whole Universe. If we deal with anything less than the whole Universe, we run into the danger of involving ourselves with a system that is wide open and to which the second law doesn't apply at all. We must always avoid making arguments involving the second law unless we are sure that our system is at least reasonably closed.

For instance, by the second law, any object which is colder than its surroundings must warm up, while the surroundings cool down until the whole system (object plus surroundings) are at equal temperature. Yet the interior of a refrigerator does *not* warm up, but remains cooler than its surroundings for an indefinite time. In fact, heat is pumped out of the refrigerator constantly so that its surroundings are warmer than they would be if the refrigerator were not there.

Does this mean that the refrigerator is violating the second law? Since it is man-made, does this mean that man is capable of violating the second law? Does this mean man can transcend natural law and has Godlike power? Or does it mean that the second law is wrong and should be discarded?

The answer to all those questions is: No!

Notice that the interior of a refrigerator begins to warm up at once when its motor is turned off. Without taking the motor into consideration, the refrigerator is simply not a closed system or anywhere near it. The motor is run by electricity that is produced by some generating system, and that, too, must be included in the system. Once that is done, it becomes clear that the entropy increase of the motor together with all that keeps it running is far higher than the entropy decrease of the refrigerator interior itself. If you take a reasonably closed system of which the refrigerator interior is part, then the second law is not violated.

Let us apply this reasoning to life itself. Life is not a closed system all by itself. Simple compounds do not spontaneously become complex compounds, or simple organisms complex ones, without something other than life itself being involved.

The compounds of the primordial sea, out of which life began, are bathed by a sea of incoming energy originating, for the most part, in the Sun (though, to a lesser degree, in the Earth's internal heat, in the radioactive substances of the Earth's crust, and so on). It is the combination of compounds *and* energy that leads to the formation and evolution of life, and *this energy must be included in the system,* if it is to be considered as reasonably closed.

Therefore, in considering the thermodynamic significance of evolution, we mustn't think of life only—for to that, the second law does not necessarily apply. We must think of the reasonably closed system of Sun and

Earth. If we do that, then we find that the entropy increase involved in the energy impinging on Earth's surface is far, far greater than the entropy decrease involved in the evolutionary changes it makes possible. In other words, the increasing order found in evolution is at the expense of a far greater increase in disorder developing in the Sun.

Evolution, therefore, once you consider it as part of a closed system (as you must) does *not* violate the second law of thermodynamics, and this particular judo argument does *not* prove the existence of God.

I am, as a matter of fact, surprised that those Believers who advance this argument (and reveal their ignorance of thermodynamics) should think the suggestion can possibly hold. Do they honestly think that scientists are so stupid that they would not see the conflict between evolution and the second law if it existed—or, seeing it, would be so lost in malice as to ignore it just to spite God?

A second judo argument goes as follows—Suppose something exists, but that the chances of its having come into existence by random processes are so small (as determined by the laws of statistics and probabilities) that it is virtually impossible to suppose that it exists except as the result of the intervention of some directing influence. Since the only directing influences we can imagine involve intelligence, and since the only form of intelligence great enough to influence major aspects of the Universe is God, we must conclude that God exists.

This argument can be advanced in general terms by saying something like: "If you grant the existence of a watch, you must assume the existence of a watchmaker, since it is impossible to believe that the delicate mechanism of a watch came to be through the fortuitous concatenation of atoms. How much more then, if we grant the existence of a Universe, must we assume the existence of a Universe-maker, who can only be God."

A more sophisticated form of the argument was presented by a French biophysicist, Pierre Lecomte du

Noüy in a book named *Human Destiny*, published in 1947, the year he died. Lecomte du Noüy calculated the chances that the various atoms making up a typical protein molecule would manage to orient themselves in just the proper fashion by chance alone. Clearly the chance of a single protein molecule forming by chance, even in the entire lifetime of the Universe, is negligible. From the fact that protein molecules nevertheless exist, in enormous numbers and great diversity, we must conclude that God exists.

I first learned of this argument ten years after it was advanced and, of course, saw the flaw in the reasoning at once. I pointed out the flaw in an article entitled "The Unblind Workings of Chance."*

Suppose, I said, we imagine not a complex protein molecule, but a very simple water molecule, consisting of two hydrogen atoms and one oxygen atom in the following order; H-O-H. Given a quantity of oxygen atoms and hydrogen atoms, we can imagine them grouping themselves into threes at random. They might arrange themselves in any of eight different combinations; OOO, OOH, OHO, HOO, OHH, HOH, HHO, HHH.

Once they have done so, you pick out one molecule at random. The chance that it is HOH is 1 in 8. The chance that the first twenty molecules you pick out at random are *all* HOH is 1 in 8^{20} or less than one out of a billion billion (10^{18}). The chances are far, far less if you also allow combinations of two atoms and four and five and so on—which might also come to pass in the kind of random assortment we are postulating.

And yet, in actual fact, if you start picking molecules out of a container in which atoms of oxygen have combined with atoms of hydrogen, we find that *all* the combinations, with negligible exceptions, are HOH.

What has happened to the laws of statistics? What has happened to randomness?

The answer is that Lecomte du Noüy, in his eager-

<hr>

* In *Only a Trillion* (Abelard-Schumán, 1957).

ness to prove the existence of God, based his argument on the assumption that atoms combine in absolutely random fashion, *and they don't*. They combine randomly only within the constraints of the laws of physics and chemistry. An oxygen atom will combine with no more than two other atoms, and with a hydrogen atom *much* more easily than with another oxygen atom. A hydrogen atom will combine with no more than one other atom. Given those rules, the only combination that forms in appreciable numbers is HOH.

Arguing similarly, you might say that while the various atoms making up protein molecules would never form a protein molecule by absolute chance—they may still do so if they combine within the constraints of their physical and chemical properties. They may combine first to form simple organic acids, then amino acids, then small peptides, and finally protein.

By the time I wrote my article, this had indeed been demonstrated experimentally. In 1955, the American chemist Stanley Lloyd Miller had begun with a small quantity of a sterile mixture of simple substances that probably existed in Earth's primordial atmosphere. He supplied the energy derived from an electric spark and, in a mere week, obtained from the mixture several organic acids and, in addition, two of the amino acids that occur in protein molecules.

Since then, other experimenters, working in similar fashion, have confirmed and vastly extended Miller's findings. Some fairly complex compounds have been formed by purely random techniques. Naturally, it is reasonable to start with compounds whose formation has already been demonstrated and use them as a new starting point. Thus, in 1958, the American biochemist Sidney W. Fox heated a mixture of amino acids and obtained protein molecules (though none that were precisely identical to any known proteins in living tissue).

So Lecomte du Noüy is wrong (although I'm sure his argument is earnestly quoted by Believers to this very day). The formation of complex compounds of the kind we associate with life is *not* such a low-probability affair that we have to call on God to extricate us from the puzzle of our own existence. It is, instead, a rather high-probability and, indeed, almost inevitable event. Given Earthlike conditions, it is difficult to see how life can avoid coming to pass.

I spoke of the inevitability of life in an article which I entitled "The Inevitability of Life" but which appeared in the June 1974 issue of *Science Digest* under the editor's title of "Chemical Evidence for Life in Outer Space." (Fie!) †

I was fascinated when, in response to that article, a letter of dissent appeared in the October 1974 issue, one that produced a judo argument in favor of God's existence that was better than Lecomte du Noüy's.

The letter writer did not try to talk about forming complex molecules atom-by-atom. Presumably he was knowledgeable enough about science to know that scientists have formed pretty complex molecules in little vats of solution over short periods of a few days. (Imagine, then, what could be done in a whole ocean of compounds over a period of a hundred million years.)

The letter writer is therefore willing to assume that the primordial ocean is full of complex molecules "with ten per cent being in the form of amino acids." He calls this a generous percentage and I suspect that it is.

He then goes on to say: "Let's further assume that these molecules are combining and recombining, making new compounds at the fastest rate known to chemistry. It's easy to prove, applying the science of mathematical probabilities, that by chance, not one recognized molecule of deoxyribonucleic acid (DNA) could be formed, even over the billions of years normally assigned to the task."

Of course, one can't make DNA out of amino acids; we need nucleotides for that. Let's dismiss that, however, as a small error by someone who is not completely at home with the matter concerning which he is arguing.

_____

† The article is included, under my original title for it, in *Of Matters Great and Small* (Doubleday, 1975).

Let us suppose we start with "trinucleotides," rather complex building blocks out of which DNA is built, and which can be built up by random processes.

A DNA molecule (or what we call a "gene" in genetics) may be made up of some four hundred trinucleotides, and each of the trinucleotides can be any of sixty-four different varieties. The total number of different DNA molecules that may be built up of four hundred trinucleotides, each one of which can be any of sixty-four, is 64^{400}, which is just about 30000000000 . . . , where you must write a total of 722 zeroes!

Now let us see how many different genes are actually known and let us multiply that number as much as we can so that we have as many different molecules out of which to select that "one recognized molecule" that we must try to form by chance if we are to confound the letter writer.

The number of different genes in a human cell may be as many as twenty-five thousand. These are duplicated in every one of the fifty trillion cells of the human body, so there are only twenty-five thousand different genes in a whole human organism as well as in one cell. Let us, however, ignore this, and pretend that every cell in the human body has twenty-five thousand genes that are different from the genes in every other cell. The total number of different genes in the human body would then be 1.25×10^{18}.

Let's go on to suppose that every one of the four billion human beings alive on Earth has a completely different set of genes, so that no human gene anywhere on the planet is like any other. In that case, the total number of different human genes on Earth would be 5×10^{27}. If we assume that the total number of non-human genes on Earth is ten million times that of human genes and that they are all different too, then the total number of genes on Earth of all kinds is 5×10^{34}.

If you go on to suppose that new genes are formed every half-hour and that they are always different, and that Earth has always been as rich in life as it is now, then in the three billion years of the history of life on

Earth, the total number of different genes that would have existed would be 2.5×10^{41}. If you suppose that this has happened not only on Earth but on each of ten different planets of every one of the hundred billion stars in our Galaxy and of every one of the stars in a hundred billion other galaxies, then the total number of different genes in the Universe is 2.5×10^{63}.

This is also a large number but compared to the total number of *possible* genes, 3×10^{722}, the total number of different genes in the Universe, even after our impossibly generous mode of computation, is so small as to be virtually zero.

If, then, you take a huge mass of nucleotide triplets and have them join at random, the chance that they will form a single "recognized molecule of DNA" in the billions of years that the Universe has existed is indeed negligible, as the letter writer states.

This is a powerful judo argument, indeed. Can we rescue ourselves by saying that the trinucleotides are *not* able to join in any fashion at all, but only within certain constraints that cause them to form only the genes we know?

Alas, no! As far as we know, the trinucleotides can join in any fashion whatever.

Have we, then, finally ended with an argument that proves that God exists?

Not quite!

There is, after all, a logical flaw in the letter writer's arguments. He makes the unspoken assumption that only the "recognized molecules" of DNA have anything to do with life—but there is no reason at all to suppose that.

In the course of the evolution of living things, new genes have constantly come into being, genes of a kind that had never existed before, genes with trinucleotide combinations not hitherto encountered. These new genes were of various types from very useful to completely useless.

There is no reason to suppose that life has exhausted all the genes that are useful to life. There is no reason

235

to suppose that a gene that is useless to one species might not be useful to another, perhaps to one that is now extinct or one that has never evolved.

It may be that a large majority of all the incredible number of genes that can be formed, but have never been formed, would, if they happened to be formed by accident, prove useful and functional in some life situation, in one way or another.

We might argue that any *particular* gene has virtually zero chance of being formed in Earth's primordial ocean, but that *some* gene was certain to form. In all likelihood, it did not matter which genes were formed as long as some genes formed. The actual direction life took and the actual fact of our own existence, may depend on the chance that certain genes were formed and not others. The Earthly forms of life are, as a result, purely fortuitous and are extremely unlikely to resemble any forms of life on any other life-bearing planets—but *the fact of some form of life* is a virtual certainty and does not require the defying of the laws of probability.

The choice, then, is not between a few select genes that lead to life and an incredibly vast majority that does not. That is only the letter writer's unspoken assumption. The choice is between one group of genes that leads to life and another that leads to a somewhat different life and still another—and still another—and still another—and still another—

Once genes are formed that represent the beginnings of a very primitive form of life, a new factor enters. The genes reproduce themselves but not always exactly, so that new genes are constantly being formed, each working a little differently.

These different genes, alone and in combination, compete with each other for existence. Survival and reproduction of this one, rather than that one, may be very largely a matter of chance, but weighting that chance ever so slightly in one direction or the other may be the comparative efficiency of working of one gene as compared to another.

Differences in efficiency or "fitness" will inevitably

lead to the survival of those genes that work best in their particular environment, and that is what is meant by "evolution through natural selection."

Genes, after having been originally formed purely by chance, are then selected by blind environmental forces into a better and better fit, until after three billion years, an organism as complex and versatile as *Homo sapiens* exists. Very likely, a species equally remarkable would have been molded by three billion years of natural selection no matter what genes had been formed in the beginning by the workings of sheer chance.

Nowhere in the entire process can I see any point where the blind laws of nature definitely break down and where we are left with no alternative but to call upon God.

Naturally, there is nothing in the argument to prove that there is no God, either. Even if we were to demonstrate that, as far as we know, God is unnecessary, we have not disproven God's existence. God may be necessary at some point that we haven't properly understood, or haven't even considered. For that matter, God may exist even if there is no necessity for that existence.

However, it is a respected principle of argument that the burden of proof is upon the positive.

Therefore, if asked whether I believe in God, I suppose I must reply that as soon as incontrovertible evidence for God's existence is presented to me, I will accept it.

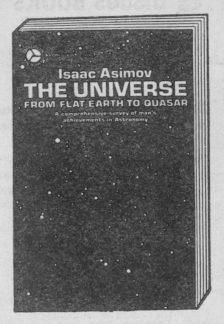

DISCUS BOOKS
DISTINGUISHED NON-FICTION

A SELECTION OF RECENT TITLES

DRT 2-78